Pitman Research Notes in Mathematics Series

Submission of proposals for consideration
Suggestions for publication, in the form of outlines and representative samples, are invited by the Editorial Board for assessment. Intending authors should approach one of the main editors or another member of the Editorial Board, citing the relevant AMS subject classifications. Alternatively, outlines may be sent directly to the publisher's offices. Refereeing is by members of the board and other mathematical authorities in the topic concerned, throughout the world.

Preparation of accepted manuscripts
On acceptance of a proposal, the publisher will supply full instructions for the preparation of manuscripts in a form suitable for direct photo-lithographic reproduction. Specially printed grid sheets are provided and a contribution is offered by the publisher towards the cost of typing. Word processor output, subject to the publisher's approval, is also acceptable.

Illustrations should be prepared by the authors, ready for direct reproduction without further improvement. The use of hand-drawn symbols should be avoided wherever possible, in order to maintain maximum clarity of the text.

The publisher will be pleased to give any guidance necessary during the preparation of a typescript, and will be happy to answer any queries.

Important note
In order to avoid later retyping, intending authors are strongly urged not to begin final preparation of a typescript before receiving the publisher's guidelines and special paper. In this way it is hoped to preserve the uniform appearance of the series.

Longman Scientific & Technical
Longman House
Burnt Mill
Harlow, Essex, UK
(tel (0279) 26721)

Titles in this series

1 Improperly posed boundary value problems
A Carasso and A P Stone
2 Lie algebras generated by finite dimensional ideals
I N Stewart
3 Bifurcation problems in nonlinear elasticity
R W Dickey
4 Partial differential equations in the complex domain
D L Colton
5 Quasilinear hyperbolic systems and waves
A Jeffrey
6 Solution of boundary value problems by the method of integral operators
D L Colton
7 Taylor expansions and catastrophes
T Poston and I N Stewart
8 Function theoretic methods in differential equations
R P Gilbert and R J Weinacht
9 Differential topology with a view to applications
D R J Chillingworth
10 Characteristic classes of foliations
H V Pittie
11 Stochastic integration and generalized martingales
A U Kussmaul
12 Zeta-functions: An introduction to algebraic geometry
A D Thomas
13 Explicit *a priori* inequalities with applications to boundary value problems
V G Sigillito
14 Nonlinear diffusion
W E Fitzgibbon III and H F Walker
15 Unsolved problems concerning lattice points
J Hammer
16 Edge-colourings of graphs
S Fiorini and R J Wilson
17 Nonlinear analysis and mechanics: Heriot-Watt Symposium Volume I
R J Knops
18 Actions of fine abelian groups
C Kosniowski
19 Closed graph theorems and webbed spaces
M De Wilde
20 Singular perturbation techniques applied to integro-differential equations
H Grabmüller
21 Retarded functional differential equations: A global point of view
S E A Mohammed
22 Multiparameter spectral theory in Hilbert space
B D Sleeman
24 Mathematical modelling techniques
R Aris
25 Singular points of smooth mappings
C G Gibson
26 Nonlinear evolution equations solvable by the spectral transform
F Calogero
27 Nonlinear analysis and mechanics: Heriot-Watt Symposium Volume II
R J Knops
28 Constructive functional analysis
D S Bridges
29 Elongational flows: Aspects of the behavi of model elasticoviscous fluids
C J S Petrie
30 Nonlinear analysis and mechanics: Heriot-Watt Symposium Volume III
R J Knops
31 Fractional calculus and integral transform generalized functions
A C McBride
32 Complex manifold techniques in theoretic physics
D E Lerner and P D Sommers
33 Hilbert's third problem: scissors congruen
C-H Sah
34 Graph theory and combinatorics
R J Wilson
35 The Tricomi equation with applications to theory of plane transonic flow
A R Manwell
36 Abstract differential equations
S D Zaidman
37 Advances in twistor theory
L P Hughston and R S Ward
38 Operator theory and functional analysis
I Erdelyi
39 Nonlinear analysis and mechanics: Heriot-Watt Symposium Volume IV
R J Knops
40 Singular systems of differential equations
S L Campbell
41 N-dimensional crystallography
R L E Schwarzenberger
42 Nonlinear partial differential equations in physical problems
D Graffi
43 Shifts and periodicity for right invertible operators
D Przeworska-Rolewicz
44 Rings with chain conditions
A W Chatters and C R Hajarnavis
45 Moduli, deformations and classifications of compact complex manifolds
D Sundararaman
46 Nonlinear problems of analysis in geometry and mechanics
M Atteia, D Bancel and I Gumowski
47 Algorithmic methods in optimal control
W A Gruver and E Sachs
48 Abstract Cauchy problems and functional differential equations
F Kappel and W Schappacher
49 Sequence spaces
W H Ruckle
50 Recent contributions to nonlinear partial differential equations
H Berestycki and H Brezis
51 Subnormal operators
J B Conway

52 Wave propagation in viscoelastic media
F Mainardi
53 Nonlinear partial differential equations and their applications: Collège de France Seminar. Volume I
H Brezis and J L Lions
54 Geometry of Coxeter groups
H Hiller
55 Cusps of Gauss mappings
T Banchoff, T Gaffney and C McCrory
56 An approach to algebraic K-theory
A J Berrick
57 Convex analysis and optimization
J-P Aubin and R B Vintner
58 Convex analysis with applications in the differentiation of convex functions
J R Giles
59 Weak and variational methods for moving boundary problems
C M Elliott and J R Ockendon
60 Nonlinear partial differential equations and their applications: Collège de France Seminar. Volume II
H Brezis and J L Lions
61 Singular systems of differential equations II
S L Campbell
62 Rates of convergence in the central limit theorem
Peter Hall
63 Solution of differential equations by means of one-parameter groups
J M Hill
64 Hankel operators on Hilbert space
S C Power
65 Schrödinger-type operators with continuous spectra
M S P Eastham and H Kalf
66 Recent applications of generalized inverses
S L Campbell
67 Riesz and Fredholm theory in Banach algebra
B A Barnes, G J Murphy, M R F Smyth and T T West
68 Evolution equations and their applications
F Kappel and W Schappacher
69 Generalized solutions of Hamilton-Jacobi equations
P L Lions
70 Nonlinear partial differential equations and their applications: Collège de France Seminar. Volume III
H Brezis and J L Lions
71 Spectral theory and wave operators for the Schrödinger equation
A M Berthier
72 Approximation of Hilbert space operators I
D A Herrero
73 Vector valued Nevanlinna Theory
H J W Ziegler
74 Instability, nonexistence and weighted energy methods in fluid dynamics and related theories
B Straughan
75 Local bifurcation and symmetry
A Vanderbauwhede
76 Clifford analysis
F Brackx, R Delanghe and F Sommen
77 Nonlinear equivalence, reduction of PDEs to ODEs and fast convergent numerical methods
E E Rosinger
78 Free boundary problems, theory and applications. Volume I
A Fasano and M Primicerio
79 Free boundary problems, theory and applications. Volume II
A Fasano and M Primicerio
80 Symplectic geometry
A Crumeyrolle and J Grifone
81 An algorithmic analysis of a communication model with retransmission of flawed messages
D M Lucantoni
82 Geometric games and their applications
W H Ruckle
83 Additive groups of rings
S Feigelstock
84 Nonlinear partial differential equations and their applications: Collège de France Seminar. Volume IV
H Brezis and J L Lions
85 Multiplicative functionals on topological algebras
T Husain
86 Hamilton-Jacobi equations in Hilbert spaces
V Barbu and G Da Prato
87 Harmonic maps with symmetry, harmonic morphisms and deformations of metrics
P Baird
88 Similarity solutions of nonlinear partial differential equations
L Dresner
89 Contributions to nonlinear partial differential equations
C Bardos, A Damlamian, J I Díaz and J Hernández
90 Banach and Hilbert spaces of vector-valued functions
J Burbea and P Masani
91 Control and observation of neutral systems
D Salamon
92 Banach bundles, Banach modules and automorphisms of C*-algebras
M J Dupré and R M Gillette
93 Nonlinear partial differential equations and their applications: Collège de France Seminar. Volume V
H Brezis and J L Lions
94 Computer algebra in applied mathematics: an introduction to MACSYMA
R H Rand
95 Advances in nonlinear waves. Volume I
L Debnath
96 FC-groups
M J Tomkinson
97 Topics in relaxation and ellipsoidal methods
M Akgül
98 Analogue of the group algebra for topological semigroups
H Dzinotyiweyi
99 Stochastic functional differential equations
S E A Mohammed

100 Optimal control of variational inequalities
V Barbu
101 Partial differential equations and dynamical systems
W E Fitzgibbon III
102 Approximation of Hilbert space operators. Volume II
C Apostol, L A Fialkow, D A Herrero and D Voiculescu
103 Nondiscrete induction and iterative processes
V Ptak and F-A Potra
104 Analytic functions – growth aspects
O P Juneja and G P Kapoor
105 Theory of Tikhonov regularization for Fredholm equations of the first kind
C W Groetsch
106 Nonlinear partial differential equations and free boundaries. Volume I
J I Díaz
107 Tight and taut immersions of manifolds
T E Cecil and P J Ryan
108 A layering method for viscous, incompressible L_p flows occupying R^n
A Douglis and E B Fabes
109 Nonlinear partial differential equations and their applications: Collège de France Seminar. Volume VI
H Brezis and J L Lions
110 Finite generalized quadrangles
S E Payne and J A Thas
111 Advances in nonlinear waves. Volume II
L Debnath
112 Topics in several complex variables
E Ramírez de Arellano and D Sundararaman
113 Differential equations, flow invariance and applications
N H Pavel
114 Geometrical combinatorics
F C Holroyd and R J Wilson
115 Generators of strongly continuous semigroups
J A van Casteren
116 Growth of algebras and Gelfand–Kirillov dimension
G R Krause and T H Lenagan
117 Theory of bases and cones
P K Kamthan and M Gupta
118 Linear groups and permutations
A R Camina and E A Whelan
119 General Wiener–Hopf factorization methods
F-O Speck
120 Free boundary problems: applications and theory, Volume III
A Bossavit, A Damlamian and M Fremond
121 Free boundary problems: applications and theory, Volume IV
A Bossavit, A Damlamian and M Fremond
122 Nonlinear partial differential equations and their applications: Collège de France Seminar. Volume VII
H Brezis and J L Lions
123 Geometric methods in operator algebras
H Araki and E G Effros
124 Infinite dimensional analysis–stochastic processes
S Albeverio
125 Ennio de Giorgi Colloquium
P Krée
126 Almost-periodic functions in abstract space
S Zaidman
127 Nonlinear variational problems
A Marino, L Modica, S Spagnolo and M Degiovanni
128 Second-order systems of partial differential equations in the plane
L K Hua, W Lin and C-Q Wu
129 Asymptotics of high-order ordinary differe equations
R B Paris and A D Wood
130 Stochastic differential equations
R Wu
131 Differential geometry
L A Cordero
132 Nonlinear differential equations
J K Hale and P Martinez-Amores
133 Approximation theory and applications
S P Singh
134 Near-rings and their links with groups
J D P Meldrum
135 Estimating eigenvalues with *a posteriori/a p* inequalities
J R Kuttler and V G Sigillito
136 Regular semigroups as extensions
F J Pastijn and M Petrich
137 Representations of rank one Lie groups
D H Collingwood
138 Fractional calculus
G F Roach and A C McBride
139 Hamilton's principle in continuum mechanics
A Bedford
140 Numerical analysis
D F Griffiths and G A Watson
141 Semigroups, theory and applications. Volu
H Brezis, M G Crandall and F Kappel
142 Distribution theorems of L-functions
D Joyner
143 Recent developments in structured continu
D De Kee and P Kaloni
144 Functional analysis and two-point differenti operators
J Locker
145 Numerical methods for partial differential equations
S I Hariharan and T H Moulden
146 Completely bounded maps and dilations
V I Paulsen
147 Harmonic analysis on the Heisenberg nilpo Lie group
W Schempp
148 Contributions to modern calculus of variati
L Cesari
149 Nonlinear parabolic equations: qualitative properties of solutions
L Boccardo and A Tesei
150 From local times to global geometry, contr physics
K D Elworthy

151 A stochastic maximum principle for optimal control of diffusions
U G Haussmann
152 Semigroups, theory and applications. Volume II
H Brezis, M G Crandall and F Kappel
153 A general theory of integration in function spaces
P Muldowney
154 Oakland Conference on partial differential equations and applied mathematics
L R Bragg and J W Dettman
155 Contributions to nonlinear partial differential equations. Volume II
J I Díaz and P L Lions
156 Semigroups of linear operators: an introduction
A C McBride
157 Ordinary and partial differential equations
B D Sleeman and R J Jarvis
158 Hyperbolic equations
F Colombini and M K V Murthy
159 Linear topologies on a ring: an overview
J S Golan
160 Dynamical systems and bifurcation theory
M I Camacho, M J Pacifico and F Takens
161 Branched coverings and algebraic functions
M Namba
162 Perturbation bounds for matrix eigenvalues
R Bhatia
163 Defect minimization in operator equations: theory and applications
R Reemtsen
164 Multidimensional Brownian excursions and potential theory
K Burdzy
165 Viscosity solutions and optimal control
R J Elliott
166 Nonlinear partial differential equations and their applications. Collège de France Seminar. Volume VIII
H Brezis and J L Lions
167 Theory and applications of inverse problems
H Haario
168 Energy stability and convection
G P Galdi and B Straughan
169 Additive groups of rings. Volume II
S Feigelstock
170 Numerical analysis 1987
D F Griffiths and G A Watson
171 Surveys of some recent results in operator theory. Volume I
J B Conway and B B Morrel
172 Amenable Banach algebras
J-P Pier
173 Pseudo-orbits of contact forms
A Bahri
174 Poisson algebras and Poisson manifolds
K H Bhaskara and K Viswanath
175 Maximum principles and eigenvalue problems in partial differential equations
P W Schaefer
176 Mathematical analysis of nonlinear, dynamic processes
K U Grusa
177 Cordes' two-parameter spectral representation theory
D F McGhee and R H Picard
178 Equivariant K-theory for proper actions
N C Phillips
179 Elliptic operators, topology and asymptotic methods
J Roe
180 Nonlinear evolution equations
J K Engelbrecht, V E Fridman and E N Pelinovski

J K Engelbrecht, V E Fridman
& E N Pelinovski

Institute of Cybernetics, Estonian Academy of Sciences

Nonlinear evolution equations

Edited by A Jeffrey, University of Newcastle upon Tyne

Longman Scientific & Technical

Copublished in the United States with
John Wiley & Sons, Inc., New York

Longman Scientific & Technical
Longman Group UK Limited
Longman House, Burnt Mill, Harlow
Essex CM20 2JE, England
and Associated Companies throughout the world.

Copublished in the United States of America with
John Wiley & Sons, Inc., 605 Third Avenue, New York, NY 10158

Originally published in Russian as НЕЛИНЕЙНЫЕ ЭВОЛЮЦИОННЫЕ УРАВНЕНИЯ
(Nonlinear evolution equations)

English language edition first published by
Longman Group UK Limited, 1988

ISSN 0269-3674

British Library Cataloguing in Publication Data
Engelbrecht, Turiĭ K.
Nonlinear evolution equations.
1. Nonlinear evolution equations
I. Title II. Fridman, V.E. III. Pelinovski,
E.N. IV. Jeffrey, Alan
515.3′53
ISBN 0-582-01314-3

Library of Congress Cataloging-in-Publication Data
Pelinovskiĭ, E.N.
[Nelineĭnye ėvoli͡utsionnye uravneniia. English]
Nonlinear evolution equations/J.K. Engelbrecht, V.E. Fridman &
E.N. Pelinovski; edited by A. Jeffrey.
p. cm.--(Pitman research notes in mathematics series, ISSN
0269-3674 ; 180)
Translation of: Nelineĭnye ėvoli͡utsionnye uravnenii͡a.
Bibliography: p.
ISBN 0-470-21148-2 (USA only)
1. Nonlinear waves. 2. Evolution equations, Nonlinear.
I. Ėngel ‘brekht, I͡Uriĭ K. II. Fridman, V.E. II. Jeffrey, Alan.
IV. Title. V. Series.
QA927.P4513 1988
515.3′53—dc19

88-9314
CIP

Printed and bound in Great Britain by
Biddles Ltd, Guildford and King’s Lynn

Contents

Abstract

Preface to the English edition

Introduction 1

1 The singular methods of simplification for one-dimensional wave processes 10

1.1 The iterative method 10

1.2 The asymptotic method 16

1.3 Asymptotic series for interacting waves 23

1.4 The spectral method 25

1.5 Strongly nonlinear systems 31

1.6 The comparative estimation of accuracy 33

2 The quasiplane waves in nonlinear media and the evolution equations 41

2.1 The asymptotic scheme for propagation along linear rays 41

2.2 The asymptotic scheme for propagation along nonlinear rays 47

2.3 The estimation of nonlinear effects in a near-caustic zone 56

2.4 The asymptotic scheme for wave-beams 63

2.5 Wave-beams in weakly inhomogeneous media 67

3 The wave-guides and evolution equations 70

3.1 The Galerkin procedure for eliminating the "non-wave" coordinate 70

3.2 The asymptotic method for the one-wave approximation 77

3.3 Wave-guides of complicated structures 88

4 Applications. Simple evolution equations 96

4.1 The equation of a simple wave 96

4.2 The Burgers equation 105

4.3 An integro-differential equation 108

4.4 The equation of nonlinear rays 110

References 114

Abstract

This work reflects contemporary understanding in the modelling of nonlinear wave processes in weakly dispersive media. The fruitful notion of evolution equations governing the propagation of single waves is used. Three methods are described for the construction of nonlinear evolution equations: the iterative, the asymptotic and the spectral. A comparative analysis of these methods is presented including the problems of the convergence and correctness. Several physical situations are discussed involving one-dimensional, weakly inhomogeneous and/or wave-guide systems. The simple evolution equations are analysed separately.

This book has been written from the viewpoint of graduate students in wave mechanics and mathematical physics and may also be of interest to post-graduate students and research workers in those branches of mathematical physics and engineering that are concerned with nonlinear wave propagation.

Preface to the English edition

Mathematical physics requires good definitions on which mathematical studies can be based. To us, one of the important notions in wave theory is the concept of single waves. Starting from this point, we return to the parts of mathematics in which the notion of evolution equations has played a significant role in the contemporary understanding on nonlinear waves. These research notes describe the methods available for constructing the evolution equations governing nonlinear wave propagation. After introducing the possible methods together with their comparative analyses, these methods are then applied to various complicated physical situations. The cases considered include wave-beams, the description of near-caustic zones, the elimination of "non-wave"coordinates, etc. The notes are written mainly on the basis of the research carried on by the authors themselves. Their scientific interests are related to nonlinear wave propagation, with applications in hydrodynamics, acoustics, mechanics, biophysics, etc. The material given in the notes has also been used for teaching at both the graduate and post-graduate level.

The authors, whose names are listed in the alphabetical order, share equal responsibility for these notes.

The translation of the notes was finished when one of the authors (JKE) was at the University of Newcastle upon Tyne, U.K., supported by a grant from the SERC. The authors are very much indebted to the SERC and would like to thank Professor Alan Jeffrey for the help in improving the presentation of the notes.

JKE, VEF, ENP

Introduction

Wave propagation theory is historically most closely related to the development of approximate methods for solving partial differential and/or integro-differential equations (systems of equations). However, even in the linear approach when finite deformations are neglected, the number of exact solutions describing the dependence of the field variables on the initial conditions is rather small. These cases, especially when three-dimensional diffraction problems and waves in dispersive media are considered, represent nowadays the classical examples of mathematical physics. The number of exactly solvable nonlinear problems is certainly much smaller. This is the main reason why approximate methods are so intensely developed for both linear and nonlinear problems. Generally speaking, the approximate methods used in the wave propagation theory may be divided into three main groups:

i) the approximate analysis of the exact solution;

ii) the perturbative analysis of the solution with small (slow) derivation from a known one;

iii) the simplification of mathematical models (equations) describing the process.

The methods of group (i) take the exact solution written in the integral form as a basis, and the approximate solutions are obtained by means of some classical approximation method, for example, by means of the methods of stationary phase and steepest descent [114, 116]. The method of stationary phase, first proposed by Kelvin for solving the wave pattern formation behind a moving ship is now considered to be a classical one. A great many linear wave propagation problems may be solved nowadays by means of such an approach when the integral solutions are known. Unfortunately, it is practically impossible to get explicit integral formulae for nonlinear problems and this is a great drawback of this physically well grounded method.

The methods of the group (ii) are better in this sense because nonlinear problems are also solvable. The basic solutions are usually taken in the form of stationary plane waves. First of all, the traditional perturbation methods belong to this group. The straightforward perturbation procedure involves a series expansion with respect to a small parameter ε, where the first term is the solution of the problem with $\varepsilon = 0$ [73, 112]. The solution of the "reduced" problem ($\varepsilon = 0$) may be found either exactly or approximately. The instability of nonlinear waves is usually analysed according to such an approach. However, as a rule, the straightforward perturbation procedures break down in the course of time t (or the space coordinate) due to secular terms which make the solution unbounded as t tends to infinity. These difficulties can be avoided if the constant parameters of the basic solution, such as the amplitude and the frequency, for example, are considered to be slowly changing in time and in space. This permits the construction of uniformly valid approximate series (or their finite sums) representing the solution at a large time. The methods based on such procedures are called singular perturbation methods [46]. The reader can turn for details to various works on the topic [2, 20, 35, 44, 108, 116].

The methods of the group (iii) do not simplify the solutions, but rather the equations governing the wave process. At this stage no attention is paid to the solutions themselves. It is clear that the simplifying procedure ought to make use of certain small parameters which may either be present in the initial equations (systems of equations) or result from the process (the solution is close to the stationary one, for example). The wave process is thus described by the solution of the simplified equations [25]. The best results here are achieved when the initial system is simplified into a single equation, first order with respect to time and of arbitrary order with respect to space coordinates. This equation is called an *evolution equation*. Physically it means that the wave process is separated into single waves, each of them described by its own equation - an evolution equation. The best example of such an evolution equation is the well-known Korteweg-de Vries equation. Its derivation and history form a brilliant chapter in contemporary mathematical physics [62].

In this book we will be concerned with the methods for constructing nonlinear evolution equations mainly in systems with the weak dispersion. Dispersion, as usual, means the dependence of the phase velocity on the frequency. In spite of weak dispersion the interaction between the different spectral components may be strong and the wave profiles may be of various shapes - from quasiharmonic waves up to N-waves and pulse-type waves. In this case it is rather difficult to fix a definite profile beforehand, therefore the most appropriate approach involves the simplification of the initial system without any concrete definition of the wave profile. The basic idea of simplification may be explained with the simple example of the linear wave equation

$$\frac{\partial^2 U}{\partial t^2} - c^2 \frac{\partial^2 U}{\partial x^2} = 0. \qquad (0.1)$$

Its well-known general solution is represented by a super-position of two waves u and v:

$$U = u(x - ct) + v(x + ct). \qquad (0.2)$$

Each single wave, for example u, satisfies a simpler equation

$$\frac{\partial u}{\partial t} + c \frac{\partial u}{\partial x} = 0. \qquad (0.3)$$

This is the simplest "single-wave" equation. Its solution coincides exactly with the solution of the initial equation (0.1) only when $v \equiv 0$, i.e. the wave process contains only one wave. The existence of a single wave is associated with certain initial and boundary conditions that actually may often occur in physics. This situation is in fact used in all the methods of simplification to a considerable extent. We shall demonstrate this approach on the physical level of strictness in connection with the model equation

$$\frac{\partial^2 U}{\partial t^2} - c^2 \frac{\partial^2 U}{\partial x^2} - \varepsilon \frac{\partial^2 U^2}{\partial x^2} - 2\,\varepsilon \frac{\partial^4 U}{\partial x^4} = 0, \qquad (0.4)$$

where $\varepsilon \to 0$. Although the solution to the linearized equation (0.4) in the form of the sinusoidal progressive wavetrains is known [46], we shall not use it further. Provided $\varepsilon = 0$ the solution to the equation (0.4) is given by expression (0.2), and, if the single wave is realized then U is a function of one variable $y = x - ct$ only. In case of small ε's the single wave solution must be given in the form $U(x,t) = u(y,\tau = \varepsilon t)$. Substituting y and τ into the equation (0.4) we obtain

$$\varepsilon^2 \frac{\partial^2 u}{\partial \tau^2} - 2\varepsilon c \frac{\partial^2 u}{\partial \tau \partial y} - \varepsilon \frac{\partial^2 u^2}{\partial y^2} - 2\varepsilon \frac{\partial^4 u}{\partial y^4} = 0. \qquad (0.5)$$

The lowest-order variation leads to

$$c \frac{\partial u}{\partial \tau} + u \frac{\partial u}{\partial y} + \frac{\partial^3 u}{\partial y^3} = 0 \qquad (0.6)$$

which is the well-known Korteweg-de Vries equation.

As it is seen, the idea of simplification is rather simple itself, although its realization for complicated physical situations described by high-order equations (systems of equations) may be often difficult and cumbersome. In this connection a vast number of papers has been published in the sixties, dealing with the derivation of evolution equations in various physical situations. The mathematical side of these problems was often very similar, and as a matter of fact, the Korteweg-de Vries equations was often the main result of these investigations. The unified approach to constructing evolution equations of the high order was suggested by Taniuti and Wei [106] and actually marks a certain milestone in the theory. By making use of the fact that the initial system governing the wave propagation is close to the hyperbolic system the authors developed an asymptotic procedure in order to simplify the system. The lowest-order approximation of this procedure gave the evolution equations. Further investigations by the authors, their co-workers and others dealt with general expressions for the coefficients of the evolution equations, with the structure of the evolution equations depending on the structure of the initial system and higher approximations etc. Nowadays we have at our disposal a solid basis in order to understand the physical mechanism of single wave processes and further construction of evolution equations with the necessary exactness in many interesting problems of physics.

The existence of single waves is not the only condition used in methods of simplification. For quasi-one-dimensional waves (elastic waves in rods, surface waves etc.) the situation involving the transversal structure of the field (of a certain mode) is fixed in a large interval of frequency and/or wave numbers. In this case it is natural to eliminate the transversal (non-wave) coordinate along which the field structure is fixed. Such an approach permits a decrease in the number of independent variables. The best example here is the "classical" derivation of the equations governing the waves in shallow water [49]. We now give a brief derivation of this result. As is well known, the movement of an ideal liquid is governed by the equation

$$\Delta\varphi = 0 \quad (-H \leqq z \leqq \eta) \tag{0.7}$$

subject to the nonlinear boundary conditions at the free surface

$$\frac{\partial\eta}{\partial t} + \nabla\varphi\nabla\eta = \frac{\partial\varphi}{\partial z}, \tag{0.8}$$

$$\frac{\partial\varphi}{\partial t} + \frac{1}{2}(\nabla\varphi)^2 + g\eta = 0, \tag{0.9}$$

and at the bottom $(z = -H)$

$$\frac{\partial\varphi}{\partial z} = 0. \tag{0.10}$$

Here φ is the velocity potential, $\eta(x,y,t)$ is the displacement of the free surface, H is the depth of the basin, g is the acceleration of gravity and Δ is the Laplacian.

The difficulties in solving this problem are obvious. Meanwhile from the solution of the corresponding linear problem (see, for example, [54]) it is known that in the long-wave limit the solution is arranged comparatively simply: the pressure is hydrostatic, the field of horizontal velocities does not depend on the depth and the vertical velocity is considerably smaller than the horizontal velocity. This is true for waves with a wavelength considerably greater than the depth of the basin. Hence the depth may be considered as a small parameter. In this case the velocity potential may be expanded into Taylor series with respect to the depth

$$\varphi = \sum_{n=0}^{\infty} \Phi_n(x,y,t)(H + z)^n. \tag{0.11}$$

By virtue of the condition (0.10) we have $\Phi_1 \equiv 0$ and the recurrence formulae

$$\Phi_{2n+1} = 0, \quad \Phi_{2n+2} = -\frac{\Delta\Phi_{2n}}{(2n + 2)(2n + 1)} \tag{0.12}$$

then follow from (0.7). Consequently, the only independent function is Φ_0. Substituting (0.11) into the boundary conditions (0.8) and (0.9) and taking (0.12) into account, we obtain the system

$$\frac{\partial\eta}{\partial t} + \nabla\eta\nabla\Phi_0 + (H + \eta)\Delta\Phi_0 - \frac{1}{6}(H + \eta)^3\Delta\Delta\Phi_0 =$$

$$= \frac{1}{2}(H + \eta)^2\nabla\eta\nabla\Delta\Phi_0 + \ldots, \tag{0.13}$$

$$g\eta + \frac{\partial\Phi_0}{\partial t} + \frac{1}{2}(\nabla\phi_0)^2 - \frac{(H + \eta)^2}{2}\frac{\partial}{\partial t}\Delta\Phi_0 =$$

$$= -\frac{(H + \eta)^2}{2}\nabla\Phi_0\nabla\Delta\Phi_0 + \ldots, \tag{0.14}$$

where the terms containing $(H + \eta)$ in fourth and higher powers are not written out. The system (0.13), (0.14) does not include the vertical coördinate z and derivatives with respect to z. The system is exact (provided all terms are taken into account) indicating that the elimination of the nonwave coordinate may be done in principle for a rather general case. It is clear, however, that the system with an infinite number of terms is not an essential simplification of the initial system. Nevertheless, if the depth H is considerably smaller than the wavelength λ, i.e. $\mu = H^2\lambda^{-2} \ll 1$ and the amplitude η is smaller than the depth, i.e. $\varepsilon = \eta H^{-1} \ll 1$ then assuming $\varepsilon \sim \mu$, all the terms on the right-hand sides are proportional to ε^2, μ^2, $\varepsilon\mu$ or to higher powers of these parameters. Neglecting these terms and introducing the particle velocity $\vec{u} = \nabla\Phi_0 - \frac{1}{2}H^2\nabla\Delta\Phi_0$ we obtain, after a certain transformation, the following system

$$\frac{\partial\vec{u}}{\partial t} + (\vec{u}\nabla)\vec{u} + g\nabla\eta = 0, \tag{0.15a}$$

$$\frac{\partial\eta}{\partial t} + \operatorname{div}\left\{(H + \eta)\vec{u} + \frac{1}{3}H^3\Delta\vec{u}\right\} = 0. \tag{0.15b}$$

These equations are called the Boussinesq equations and they generalize the equations of shallow water where the dispersion is not taken into account. Thus the standard expansion of the velocity potential into a series with the respect to the nonwave coordinate permits the construction of much simpler equations. Certainly, depending on the structure of the initial equations the formal procedures may be different, and nowadays they are extensively used for many waveguide problems.

Finally, let us demonstrate one approach more thoroughly in order to get the evolution equation. This approach is based on the dispersion relation [116]

$$\omega = f(k), \tag{0.16}$$

where ω, generally speaking, is the complex frequency and k is the real wave number. If this relation is multiplied by $iu(k)\exp\{i(\omega t - kx)\}$, where $u(k)$ is an arbitrary function and i is the imaginary unit, then after integration with respect to k, we obtain

$$\frac{\partial u}{\partial t} = \hat{F}(u) = \int if(k)u(k)\exp\{i(\omega t - kx)\}dk. \tag{0.17}$$

Here $u(x,t)$ and $u(k)$ are linked by the Fourier transform

$$u(x,t) = \int u(k)\exp\{i(\omega t - kx)\}dk. \tag{0.18}$$

The expression (0.17) is a fundamental one determining the structure of the evolution equation in terms of the dispersion relation (0.16). If the function $f(k)$ is a polynomial, then the operator $\hat{F}$ is differential. In particular, if

$$f(k) = ck - \beta k^3 + i\delta k^2, \tag{0.19}$$

then equation (0.17) takes the following form

$$\frac{\partial u}{\partial t} + c\frac{\partial u}{\partial x} + \beta\frac{\partial^3 u}{\partial x^3} - \delta\frac{\partial^2 u}{\partial x^2} = 0, \tag{0.20}$$

i.e. the expression (0.19) corresponds to the linear Korteweg-de Vries-Burges equation.

The polynomial dependence of the frequency ω on the wave-number k corresponds to systems governed by evolution equations of the differential type. Contrary to this, the dispersion relations for many boundary problems are of transcendental type that lead to the evolution equation of integro-differential form. We place emphasis on the fact that $\omega(k)$ may be determined not only theoretically but also experimentally when the phase velocity and the decrement are known. This "experimental" approach is rather important for physical applications, in particular for geophysics, since the velocities of seismic waves of several types are determined only experimentally. However, even when the initial equations are known and $\omega(k)$ is determined theoretically, the simple approach described above has certain advantages. The main point here is the possibility of checking the derivation of the evolution equations carried out by asymptotic methods. This is essential in the case of nonlinear problems when the linear part of the evolution equation may be obtained from equation (0.17). Actually it means predicting the form of the evolution equations, which may help in choosing between possible analytical methods in order to find an optimal one.

The three examples given above demonstrate the effectivness of three different *methods* developed for the construction of evolution *equations*. The variety of methods and essential differences in physical problems and/or mathematical description determine the structure of this book. Three chapters deal with the methods of simplification for various physical situations including one-dimensional waves, weakly inhomogeneous media and waveguides. The fourth chapter deals with specific evolution equations. On the one hand, we have been trying to reach maximal generality and exactness in the formalisation of mathematical algorithms, while on the other hand we have tried to demonstrate the applicability of the derived methods to specific physical problems. The problems investigated in this book cover practically all aspects of wave dynamics in homogeneous, weakly inhomogeneous and stratified media.

No attempt has been made at completeness of the reference list; however, the selection gives a clear account of the topics under consideration. There are several well-known monographs [24, 46, 73, 116] that are close to the subject matter of this book, dealing with one-dimensional waves and weakly inhomogeneous media. In this book, however, all methods of simplification are presented from the unified viewpoint of the wave processes in media with arbitrary dimensionality.

The authors hope that the book will be of use to research workers concerned with nonlinear wave propagation in weakly dispersive media and may help them to simplify complicated physical problems.

We would like to express our gratitude to V.N. Goldberg, K.A. Gorschkov, M.E. Kutser, and L.A. Ostrovsky for helpful advice and comments. Our special thanks go to V.I. Schrira for his constructive criticism while acting as Editor to the Russian edition of the book.

1 The singular methods of simplification for one-dimensional wave processes

The physics of one-dimensional wave processes has been thoroughly investigate and therefore the asymptotic methods for constructing the one-dimensional evolution equations have been elaborated in great detail. In this chapter, however, most attention will be paid to the mathematical side of the process. An attempt is made to present three methods for constructing evolution equations from a unified viewpoint. Therefore the main emphasis is on comparing the different methods and analysing the structure of evolution equations of various orders.

1.1 The iterative method

The iterative method, it seems, leads to a physically well-grounded sequence of evolution equations [82]. Let the one-dimensional wave process be governed by a system of equations written in the following matrix form

$$I \frac{\partial \vec{U}}{\partial t} + A(\tau,\chi) \frac{\partial \vec{U}}{\partial x} = \hat{\vec{F}} \{\vec{U},x,t,\chi,\tau,\varepsilon\}, \qquad (1.1a)$$

$$\tau = \varepsilon t, \ \chi = \varepsilon x, \quad \varepsilon << 1, \qquad (1.1b)$$

where $\vec{U}$ is an N-vector of field variables, I is the unit matrix, A is an $(N \times N)$ matrix and $\hat{\vec{F}}$ is a given nonlinear integro-differential operator. We assume that all of the N eigenvalues $\lambda_1, \lambda_2, \ldots, \lambda_N$ of the matrix A are real and distinct, i.e. if $\varepsilon = 0$ then the system (1.1a) is totally hyperbolic.

The system (1.1a) must be solved subject to boundary and the initial conditions. The number of the conditions depends on the order of the system (i.e. on the order of the derivatives and integrals in the operator $\hat{\vec{F}}$). The simplification of the boundary and initial conditions is carried out in the same way as the simplification of the initial system so this process will not further be analysed in detail. In the case $\varepsilon = 0$ system (1.1a) has a solution in the form of the following set of waves:

$$\vec{U} = \sum_{i=1}^{N} v_i(x - \lambda_i t)\vec{r}_i, \tag{1.2}$$

where v_i is an arbitrary scalar function determined from the boundary and initial conditions, and $\vec{r}_i$ is the right eigenvector of the matrix A corresponding to the eigenvalue (velocity) λ_i. Thus we have

$$A\vec{r}_i = \lambda_i\vec{r}_i, \ \mathrm{Det}|A - \lambda_i\vec{r}| = 0. \tag{1.3}$$

In order to investigate a large class of perturbed waves ($\varepsilon \neq 0$) we shall transform the initial system (1.1a) into the normal form [17]. Let us introduce the new variable $\vec{V}$ determined by

$$\vec{U} = Y\vec{V}, \tag{1.4}$$

where Y is an ($N \times N$) matrix composed of the linearly independent eigenvectors of the matrix A. The eigenvectors $\vec{r}$ and the velocities λ depend on the "slow" variables τ and χ since the matrix A involves the "slow" time and coordinate. Substituting (1.4) into the equation (1.1a) we obtain the system of equations

$$\frac{\partial\vec{V}}{\partial t} + \Theta\frac{\partial\vec{V}}{\partial x} = \varepsilon Y^{-1}\ \{\hat{\vec{F}} - (\frac{\partial Y}{\partial\tau} + A\ \frac{\partial Y}{\partial\chi})\vec{V} =$$

$$= \varepsilon\hat{\vec{f}}(\vec{V},x,t,\tau,\chi,\varepsilon\}, \tag{1.5}$$

where Θ is a diagonal matrix with the elements $\lambda_i(\tau,\chi)$ and Y^{-1} is the inverse matrix to Y, composed of the left eigenvectors $\vec{\ell}$ of the matrix A. Here and elsewhere we assume the normalisation of eigenvectors

$$\vec{\ell}_i.\vec{r}_j = \delta_{ij}\ , \tag{1.6}$$

where δ_{ij} is the Kronecker symbol.

Writing the initial system in the form (1.5) we are able to bring about an essential simplification of the procedure for constructing the successive approximations with respect to the powers of the small parameter ε. Physically the idea is clear. According to the expression (1.4) and the system (1.5) the desired function is the superposition of waves, each of them (at $\varepsilon = 0$) is propagating along its characteristics in the x,t plane and is interacting with the other waves only because of the perturbation ($\vec{f} \neq 0$). In many cases, boundary and initial conditions are such that at $\varepsilon = 0$ only $m < N$ waves or even only one of them may occur. Another example is a localized initial disturbance which decays into separate pulses, each diverging along their respective characteristics so that at large t each of them may be considered independently. In this case we may single out $m < N$ of the main variables (components of the vector $\vec{V}$), with the rest of them $(N - m)$ being small. Physically this means that we have m main waves while the other $N - m$ waves must be determined by means of a perturbation method. As a result, we obtain the following approximations:

zero-approximation (ε^0)

$$\frac{\partial v_i^{(0)}}{\partial t} + \lambda_i(\chi,\tau) \frac{\partial v_i^{(0)}}{\partial x} = 0, \; i = 1,\ldots,m, \quad (1.7)$$

$$v_s^{(0)} = 0, \; i = m + 1,\ldots,N; \quad (1.8)$$

the first approximation (ε^1)

$$\frac{\partial v_i^{(0)}}{\partial t} + \lambda_i(\chi,\tau) \frac{\partial v_i^{(1)}}{\partial x} =$$

$$= \varepsilon f_i\{v_1^{(1)},\ldots,v_M^{(1)}, 0,\ldots,0; \chi,\tau,0\}, \quad (1.9)$$

$$\frac{\partial v_s^{(1)}}{\partial t} + \lambda_s(\chi,\tau) \frac{\partial v_s^{(1)}}{\partial x} =$$

$$= \varepsilon f_s \{v_1^{(1)},\ldots,v_m^{(1)}, 0,\ldots,0; \chi,\tau,0\}. \quad (1.10)$$

the n-th approximation (ε^n)

$$\frac{\partial v_i^{(n)}}{\partial t} + \lambda_i(\chi,\tau)\,\frac{\partial v_i^{(n)}}{\partial x} =$$

$$= \varepsilon f_i\{(v_1^{(n)},\ldots,v_m^{(n)},\ v_{m+1}^{(n-1)},\ldots,v_N^{(n-1)};\ ,\chi,\tau\ ,\varepsilon\} \qquad (1.11)$$

$$\frac{\partial v_s^{(n)}}{\partial t} + \lambda_s(\chi,\tau)\,\frac{\partial v_s^{(n)}}{\partial x} =$$

$$= \varepsilon f_s\ \{v_1^{(n)},\ldots,v_m^{(n)},\ v_{m+1}^{(n-1)},\ldots,v_N^{(n-1)};\ ,\chi,\ \tau.\varepsilon\} \qquad (1.12)$$

Thus, at each approximation the system of equations is divided into two groups: m coupled nonlinear evolution equations (1.11) for the main waves and N - m independent linear equations (1.12) with given right-hand sides describing the small corrections to the main waves. As a result, we obtain an iteration scheme for the construction of the evolution equations of m interacting waves.

Let us consider the case of one-wave generation (m = 1). Setting $v_1^{(n)} = v$, $\lambda_1 =\ $, $f_1 = f$ we shall rewrite equation (1.11) in the form

$$\frac{\partial v}{\partial t} + \lambda(\chi,\tau)\,\frac{\partial v}{\partial x} = \varepsilon f\ v,\ v_s^{(n+1)};\ \chi,\tau,\varepsilon\}. \qquad (1.13)$$

Here in the first approximation the function f must be calculated at $\varepsilon = 0$, i.e. at $v_s = 0$. The order of the evolution equation of the first order naturally depends on the form of the functional $\hat{\vec{F}}$ in the initial system (1.1a). In many cases the functional $\hat{\vec{F}}$ may be represented as a sum of derivatives and integrals containing the field variables. Since the function f is calculated from the expression

$$f = \vec{\ell}_1\hat{\vec{F}} - \vec{\ell}_1\,\frac{\partial \vec{r}_1}{\partial \tau}\,v + \sum_{i,j=1}^{N} \ell_{1j}A_{ji}\,\frac{\partial r_{1i}}{\partial \chi}\,v, \qquad (1.14)$$

following from (1.5), the function f may also be written as a sum of certain derivatives and integrals involving the variable v. It is to be understood that, according to the expression (1.14), this sum is not equivalent to the structure of the functional $\hat{\hat{F}}$. The evolution equation of the first approximation now reduces to

$$\frac{\partial v}{\partial t} + \lambda(X,\tau)\frac{\partial v}{\partial x} + \ldots + \varepsilon\int h^{(1)}dx + \varepsilon h^{(0)} +$$

$$+ \varepsilon\frac{\partial h^{(1)}}{\partial x} + \varepsilon\frac{\partial^2 h^{(2)}}{\partial x^2} + \varepsilon\frac{\partial^3 h^{(3)}}{\partial x^3} + \ldots = 0, \qquad (1.15)$$

where $h^{(n)}$ are functions of v only and all the derivatives with respect to t are excluded with the aid of (1.13). In the simplest case only $h^{(1)}$ is non-zero and the corresponding evolution equation is

$$\frac{\partial v}{\partial t} + \lambda\frac{\partial v}{\partial x} + \varepsilon\frac{\partial h^{(1)}(v)}{\partial x} = 0. \qquad (1.16)$$

This is the equation of simple waves governing wave propagation in a nondispersive medium. If $h^{(1)}$ and $h^{(2)}$ are both non-zero, and $h^{(1)} = \frac{1}{2}v^2$, $h^{(2)} = -\delta v$, then the resulting equation is the Burgers equation

$$\frac{\partial v}{\partial t} + \lambda\frac{\partial v}{\partial x} + \varepsilon v\frac{\partial v}{\partial x} - \varepsilon\delta\frac{\partial^2 v}{\partial x^2} = 0, \qquad (1.17)$$

governing the wave propagation in a nonlinear dissipative medium. If $h^{(1)}$ and $h^{(3)}$ are non-zero and $h^{(1)} = \frac{1}{2}v^2$, $h^3 = \beta v$ then the resulting evolution equation is the Korteweg-de Vries equation

$$\frac{\partial v}{\partial t} + \lambda\frac{\partial v}{\partial x} + \varepsilon v\frac{\partial v}{\partial x} + \varepsilon\beta\frac{\partial^3 v}{\partial x^3} = 0. \qquad (1.18)$$

This governs nonlinear wave propagation in a dispersive medium. Here the dispersion is more essential for short waves (high-frequency dispersion). The equations (1.16) - (1.18) have been well investigated in the nonlinear wave propagation theory [16, 24, 31, 49, 97, 100, 110, 121], and their exact solutions subject to initial conditions (the Cauchy problem) are known.

The physical role of the terms $h^{(1)}$, $h^{(2)}$ and $h^{(3)}$ in (1.15) is clear. The other terms have the meanings we now describe. The term $h^{(0)}$ governs the inhomogeneity and/or the time-variability of the medium. It may also describe the frequency-independent losses. The term $h^{(-1)}$ corresponds to low-frequency dispersion. It should be pointed out that relaxing media are often characterized by integral terms $\int K(x - x')v(x')dx'$ in the evolution equation. Here $K(x - x')$ is a kernel function [25, 97].

Let us now discuss the structure of evolution equations of higher approximations. The order of these equations is determined by the solution of the equations (1.12) governing the small corrections v_s in all of the preceding approximations. The order of the evolution equations depends on the structure of the functional $\hat{\vec{F}}$ in a rather complicated way. Suppose first that $\hat{\vec{F}} \sim \partial\vec{U}/\partial x$, then $v_s \sim \varepsilon v$ and provided the function f is also proportional to $\partial v_s/\partial x$, the order of the equation (1.13) remains the same. Consequently, in a nondispersive medium the corrections do not change the order of the evolution equations but they define more exactly the phase velocity, by taking the nonlinear effects into account. If $\hat{\vec{F}} \sim \partial^2 U/\partial x^2$ then in the first approximation $v_s^{(1)} \sim \partial v/\partial x$ and in the evolution equation of the second approximation the term $f \sim \partial^2 v_s/\partial x^2$ gives rise to a term $\partial^3 v/\partial x^3$, i.e. the order is raised. This means that every successive approximation also raises the order of the evolution equation (1.13). In particular, such corrections to the Burgers equation modelling the processes of nonlinear acoustics were known earlier [58]. The term $\partial^3\vec{U}/\partial x^3$ raises the order of the evolution equations twice etc. If $\hat{\vec{F}} \sim \vec{U}$, then in the second approximation we get $\int vdx$, and in the third approximation double integral etc. This corresponds to all cases when the functional $\vec{F}$ contains integrals of $\vec{U}$. If, however, $\hat{\vec{F}} \sim \int \vec{U}dx$, then the multiplicity will be doubled for each following approximation. Thus, the order of the evolution equation will be changed differently in systems with high-frequency and low-frequency dispersion (dissipation). The dispersion leads to higher derivatives, the dissipation - to multiple integrals. The "boundary-layer" type case corresponds to a non-dispersive medium where the order of the evolution equations does not undergo any changes in any of the approximations. It must be mentioned that in the case of a linear medium these results are trivial and correspond to the expansion of the exact dispersion relation $\omega(k)$ into a Taylor series. It is quite natural that the accuracy might be improved by taking more terms in the series into account and every higher order of k or k^{-1} also changes the order of the evolution equations (see Introduction). Finally we note that the power of the nonlinear terms increases with each approximation. In particular, if $\vec{F} \sim \vec{U}^2$ then in the second approximation we get $f \sim v^3$ etc. These conclusions are clearly valid also in case of the more complicated system (1.11) involving many waves.

We have shown here that the right-hand side of every evolution equation of an arbitrary approximation may be transformed with the aid of (1.13) into a more informative form consisting of derivatives and/or integrals with respect to only one variable (either the time or a space coordinate). Therefore the statement of the boundary (initial) problem for equation (1.13) should be carried out similarly: only the functions v(x,0) or v(0,t) must be given. For the initial system (1.1), however, depending on the structure of the functional $\hat{\vec{F}}$, as well as the functions v(x,0) or v(0,t), derivatives of the necessary order must also be given. Since the right-hand side of system (1.1) contains the small parameter ε, these derivatives are not arbitrary but must be calculated from the system (1.1) with the appropriate accuracy.

The iterative scheme for the constructing of the evolution equations described above is both transparent physically and convenient in practice, especially for the first approximation. However, its connection with the other asymptotic expansions is not clear within the asymptotic framework presented here. Moreover, it is rather difficult to elaborate effective numerical algorithms for solving the successive sequence of the evolution equation with increasing orders in successive approximations. Therefore in Section 1.2 another scheme for constructing evolution equations is presented, based on the expansion of the solutions describing the wave fields [75, 108].

1.2 The asymptotic method

The simple asymptotic scheme. Let us consider once more the initial system (1.1). At $\varepsilon = 0$ its solution is given by the superposition of waves (1.2). If boundary and initial conditions obey the one-wave situation, then at $\varepsilon \neq 0$ it is natural to seek a solution in the following form

$$\vec{U} = \vec{U}(\xi,\tau), \qquad (1.19)$$

where in the case of a homogeneous and stationary medium we have

$$\xi = x - \lambda t, \quad \tau = \varepsilon t \tag{1.20}$$

for the initial problem and

$$\xi = x - \lambda t, \quad \tau = \varepsilon x \tag{1.21}$$

for the boundary problem. Here λ is one of the eigenvalues of the matrix A corresponding to the wave under the consideration. Such an approach always leads to an initial problem for the evolution equation [84, 116]. Let us consider first the initial problem. Substituting (1.20) into the initial system we obtain in terms of ξ and τ

$$(A - \lambda I)\frac{\partial \vec{U}}{\partial \xi} = \varepsilon\left\{\hat{\vec{F}} - I\frac{\partial \vec{U}}{\partial \tau}\right\}. \tag{1.22}$$

We shall assume here smoothness of the functional $\hat{\vec{F}}$ with respect to its arguments. The solution of the system (1.22) is now sought in the form of an asymptotic series

$$\vec{U} = \vec{U}_0 + \varepsilon\vec{U}_1 + \varepsilon^2\vec{U}_2 + \dots . \tag{1.23}$$

Introducing (1.23) into the equation (1.22) and equating the coefficients of like powers in ε, we obtain from the powers ε^0, ε^1 and ε^2

$$(A - \lambda I)\frac{\partial \vec{U}_0}{\partial \xi} = 0, \tag{1.24}$$

$$(A - \lambda I)\frac{\partial \vec{U}_1}{\partial \xi} = \hat{\vec{F}}\{\vec{U}_0, x, t, \varepsilon = 0\} - I\frac{\partial \vec{U}_0}{\partial \tau}, \tag{1.25}$$

$$(A - \lambda I)\frac{\partial \vec{U}_2}{\partial \xi} = \frac{\delta \hat{\vec{F}}}{\delta \vec{U}_0}\vec{U}_1 + \frac{\partial F}{\partial \varepsilon} - I\frac{\partial \vec{U}_1}{\partial \tau}, \tag{1.26}$$

respectively. Here the variational derivative $\delta\hat{\vec{F}}/\delta\vec{U}_0$ and $\partial\hat{\vec{F}}/\partial\varepsilon$ are determined at $\varepsilon = 0$, i.e. they contain only the solutions of the zero approximation. The structure of higher approximations is equivalent to

(1.26) and their right-hand sides include the functions from the previous approximations. Due to their rather awkward form we shall present them here only in the general form of the n-th approximation

$$(A - \lambda I) \frac{\partial \vec{U}_n}{\partial \xi} = \hat{\vec{H}}_n \left\{\vec{U}_0, \ldots, \vec{U}_{n-1}\right\} - I \frac{\partial \vec{U}_{n-1}}{\partial \tau}, \qquad (1.27)$$

where $\hat{\vec{H}}_n$ includes the successive derivatives from $\hat{\vec{F}}$.

To get the solution of the zeroth approximation is trivial. As the determinant of the matrix $A - \lambda I$ must equal to zero (see (1.3)), the solution (U_0) is determined with an accuracy up to the arbitrary scalar function $v(\xi,\tau)$

$$\vec{U}_0 = v(\xi,\tau)\vec{r}, \qquad (1.28)$$

where $\vec{r}$, as earlier, is the right eigenvector of the matrix A at fixed λ.

Due to the condition $\mathrm{Det}|A - \lambda I| = 0$ the solution of the equation of the first approximation (1.25), exists only when the orthogonality conditions

$$\vec{\ell}\left\{\hat{\vec{F}} - I \frac{\partial \vec{U}_0}{\partial \tau}\right\} = 0 \qquad (1.29)$$

are satisfied. This follows immediately from (1.25) after multiplying it from the left with the left eigenvector $\vec{\ell}$ of the matrix A. Substituting (1.28) into (1.29) the last equation becomes

$$\frac{\partial v}{\partial \tau} = \hat{f}\{V\} = \vec{\ell}\hat{\vec{F}}(v\vec{r}). \qquad (1.30)$$

As a result we have obtained the evolution equation of the first approximation with respect to the function v. It must be solved subject to the corresponding initial equations. It is clear that the equation (1.30) coincides with the equation (1.13) provided the new variables (1.20) are introduced into (1.13). In other words, the iterative and the asymptotic schemes lead to the same evolution equation in the first approximation. It should be emphasized at once that asymptotic series are preferable when compared with the usual ones if all the terms in the series (1.23) are dependent only on ξ. In this case the right-hand side of (1.25) is determined and calculated from the zeroth approximation. It means that the orthogonality conditions may not be satisfied. As a result, the solution $\vec{U}_1$ may not be bounded and then the perturbation series is not convergent. In the case of asymptotic expansions we have freedom when choosing the dependence of $\vec{U}_0$ on τ. This gives the possibility of choosing so that $\vec{U}_1$ will be bounded. Such a choice is realized by choosing $v_0(\xi,\tau)$ as the solution of the evolution equation of the first order.

In order to construct the evolution equations of the second approximation, the solution of the system (1.25) must be written in the explicit form. Satisfying the orthogonality conditions (1.29) the general solution of (1.25) may be written as

$$\vec{U}_1 = v_1(\xi,\tau)\vec{r} + \vec{U}_{10}(\xi,\tau). \tag{1.31}$$

Here $\vec{U}_{10}$ is the forced solution of the system (1.25) determined by the solution of the evolution equation (1.30), and v_1 is an arbitrary scalar function. This function may be found in similar fashion to the one described above for the first approximation. Now we use the orthogonality conditions for the solution of the second approximation $\vec{U}_2$, i.e. the right-hand side of (1.26) is orthogonal to the left eigenvector $\vec{\ell}$. Accordingly we get the evolution equation for the second approximation

$$\frac{\partial v_1}{\partial \tau} = \vec{\ell}\,\frac{\delta \vec{F}}{\delta \vec{U}_0}\,(v_1\vec{r} + \vec{U}_{10}) + \vec{\ell}\,\frac{\partial \vec{F}}{\partial \varepsilon}\,. \tag{1.32}$$

In this equation the function $\vec{U}_{10}$, determined from the solution of the evolution equation of the first approximation, is given.

Respectively, the solution of the n-th approximation is

$$\vec{U}_n = v_n(\xi,\tau)\vec{r} + \vec{U}_{n0}(\xi,\tau), \tag{1.33}$$

where $\vec{U}_{n0}$ is the forced solution of system (1.27) involving $\vec{U}_0,\ldots,\vec{U}_{n-1}$, and v_n is the scalar function satisfying the evolution equation of n-th order

$$\frac{\partial v_n}{\partial \tau} = \vec{\ell}\hat{\vec{H}}_{n-1}\{\vec{U}_0,\ldots,\vec{U}_{n-1},\, v_n\}, \tag{1.34}$$

following from the orthogonality condition of the solution of the n + 1st approximation.

Thus we have presented an asymptotic scheme for constructing the evolution equation at every approximation. In the first approximation the evolution equation is a nonlinear (and autonomous, if $\hat{\vec{F}}$ does not depend explicitly on x and t) one, the next approximations are linear but with variable coefficients.

The modified asymptotic scheme. As shown above, the evolution equations of the first approximation constructed by the iterative and asymptotic methods coincide, whereas the evolution equations of higher approximations differ. It is possible to modify the asymptotic procedure in such a way that the evolution equations obtained by this method coincide with those obtained by the iterative method in every approximation. Let us introduce the various scales into the independent variables

$$\xi = x - \lambda t, \ \tau_1 = \varepsilon t, \ \ \tau_2 = \varepsilon^2 t, \dots, \ \tau_n = \varepsilon^n t, \dots \ . \tag{1.35}$$

Substituting (1.35) into the initial system (1.1) we obtain

$$(A - \lambda I) \frac{\partial \vec{U}}{\partial \xi} = \varepsilon \hat{\vec{F}} - \sum_{n=1}^{\infty} \varepsilon^n I \frac{\partial \vec{U}}{\partial \tau_n} \ . \tag{1.36}$$

The solution to (1.36) is sought in the form of the asymptotic expansion (1.23). Similarly to (1.27) the n-th order approximation is governed by

$$(A - \lambda I) \frac{\partial \vec{U}_n}{\partial \xi} = \vec{H}_n \left\{ \vec{U}_0, \dots, \vec{U}_{n-1} \right\} - \\ - I \left\{ \frac{\partial \vec{U}_{n-1}}{\partial \tau_1} + \dots + \frac{\partial \vec{U}_0}{\partial \tau_n} \right\} . \tag{1.37}$$

The solution of the zeroth approximation is given by (1.28) where the function v depends on $\xi, \tau_1, \dots, \tau_n$. The solution of the first approximation exists provided the orthogonality conditions (1.29) are satisfied when an evolution equation of type (1.30) follows

$$\frac{\partial v}{\partial \tau_1} = \vec{\ell}\hat{\vec{F}}\{v\vec{r}\}. \tag{1.38}$$

Solving this equation, the solution to the system (1.36) at n = 1 may be represented in the form of the forced solution $\vec{U}_{10}$ and the fundamental solution $v_1\vec{r}$ similarly to the previous case (see (1.31)). Contrary to (1.31), however, there is no need to consider the fundamental solution, while a definite choice of $\partial v/\partial \tau_2$ guarantees the boundedness of the correction of the second approximation. Therefore we assume v_1, v_2,... equal to zero. Then the orthogonality condition of the right-hand side of (1.37) at n = 2 gives the evolution equation of the second approximation (c.f. (1.32))

$$\frac{\partial v}{\partial \tau_2} = \vec{\ell}\,\frac{\delta \vec{F}}{\delta \vec{U}_0}\,\vec{U}_{10} + \vec{\ell}\,\frac{\partial \vec{F}}{\partial \varepsilon} - \vec{\ell}\,\frac{\partial \vec{U}_{10}}{\partial \tau_1}\ . \tag{1.39}$$

In the same way we obtain the evolution equation of the n-th approximation (c.f. (1.34))

$$\frac{\partial v}{\partial \tau_n} = \vec{\ell}\hat{\vec{H}}\left\{v,\ \vec{U}_{10},\ldots,\vec{U}_{n-10}\right\} - \vec{\ell}\left\{\frac{\partial \vec{U}_{10}}{\partial \tau_{n-1}} + \ldots + \frac{\partial \vec{U}_{n-10}}{\partial \tau_n}\right\}\ . \tag{1.40}$$

As shown, according to the modified scheme we have obtained a sequence of equations for the function v as a function of many independent variables. Taking into account that

$$\frac{\partial v}{\partial \tau} = \frac{\partial v}{\partial \tau_1} + \varepsilon\frac{\partial v}{\partial \tau_2} + \varepsilon^2\,\frac{\partial v}{\partial \tau_3} + \ldots + \varepsilon^{n-1}\,\frac{\partial v}{\partial \tau_n} + \ldots\ , \tag{1.41}$$

this sequence turns into one equation concerning the function $v(\xi,\tau)$ of only two independent variables. It means that the iterative and asymptotic methods give the same result and the choice between them, generally speaking, depends on the preference of the researcher. A comparative analysis of the methods is given in Section 1.6 where a model example is solved.

Finally we point out that a boundary problem will be solved in exactly the same way. By making use of (1.21) and (1.23) we may write the equations of the n-th approximation in the form

$$(A - \lambda I) \frac{\partial \vec{U}_n}{\partial \xi} = \vec{H}_n - A \frac{\partial \vec{U}_{n-1}}{\partial X}, \tag{1.42}$$

and due to the boundedness of $\vec{U}_n$ the evolution equations

$$\lambda \frac{\partial v_n}{\partial X} = \vec{\ell}\hat{\vec{H}}_n \tag{1.43}$$

then follow. These equations coincide exactly with the equations (1.34) if the substitution $X - \lambda\tau$ is taken into account.

If the medium is inhomogeneous and stationary ($A = A(X)$) then, using the change of variables

$$\xi = t - \int \frac{dx}{\lambda(x)}, \quad X = \varepsilon x, \tag{1.44}$$

we obtain for every power $\varepsilon^0, \varepsilon^1, \varepsilon^2, \ldots$ equations similar to (1.35) and (1.36). If the medium is homogeneous but non-stationary then the change of variables

$$\xi = x - \int \lambda(t)dt, \quad \tau = \varepsilon t \tag{1.45}$$

must be used. Finally, if the medium is inhomogeneous and non-stationary then the space-time characteristic ξ is determined by the equation

$$\frac{\partial \xi}{\partial t} + \lambda(\tau, X) \frac{\partial \xi}{\partial x} = 0 \tag{1.46}$$

with either $\tau = \varepsilon t$ or $X = \varepsilon x$ depending on the problem under the consideration (initial and boundary problems, respectively). Further analysis should be carried out in similar fashion.

1.3 Asymptotic series for interacting waves

The asymptotic method for constructing the evolution equations presented in the previous Section may be generalized for the case of interaction of several waves [24, 76, 109]. Here we start with the asymptotic expansion

$$\vec{U} = \vec{U}_0 + \varepsilon\vec{U}_1 + \varepsilon^2\vec{U}_2 + \dots . \tag{1.47}$$

Substituting (1.47) into the initial system (1.1) and equating the coefficients of like powers in ε, we obtain the sequence of approximate equations. The equations of the zeroth approximation are

$$I\frac{\partial\vec{U}_0}{\partial t} + A\frac{\partial\vec{U}_0}{\partial x} = 0. \tag{1.48}$$

The solution of the above system is

$$\vec{U}_0 = \sum_{i=1}^{m} v_i(x - \lambda_i t)\vec{r}_i, \tag{1.49}$$

where m is the number of interacting waves, and v_i is a scalar function describing the wave that propagates with the velocity λ_i. Such a form stipulates using several independent "moving" variables at $\varepsilon \neq 0$

$$\xi_i = x - \lambda_i t, \ \tau = \varepsilon t, \quad i = 1,\dots,m. \tag{1.50}$$

The solution (1.49) in terms of the new variables (1.50) may be rewritten as

$$\vec{U}_0 = \sum_{i=1}^{m} v_i(\xi_i,\tau,\zeta_i)\vec{r}_i, \ \zeta_{j\neq i} = \varepsilon\xi_j. \tag{1.51}$$

The freedom in choosing the dependence of v on τ allows the possibility of constructing an asymptotic series without secular terms. The equations of the first approximation in terms of (1.51) are the following

$$I \frac{\partial \vec{U}_1}{\partial t} + A \frac{\partial \vec{U}_1}{\partial x} = \hat{\vec{F}}\{x,t,\vec{U}_0\} - \sum_{i=1}^{m} \frac{\partial v_i}{\partial \tau} \vec{r}_i -$$

$$- \sum_{i=1}^{m} \sum_{j=1}^{m} (\lambda_i - \lambda_j) \vec{r}_i \frac{\partial v_i}{\partial \zeta_j} . \qquad (1.52)$$

In order to get conditions about the boundedness of the solutions of the first approximation we shall transform the system (1.52) into its normal form by changing to the variables

$$\vec{U}_1 = Y\vec{V}_1. \qquad (1.53)$$

Similarly to Section 1.1, Y is an $(N \times N)$ matrix composed of the linearly independent eigenvectors of the matrix A. Substituting (1.53) into (1.52) we obtain the system of decomposed equations

$$\frac{\partial V_{1i}}{\partial t} + \lambda_i \frac{\partial V_{1i}}{\partial x} = \vec{\ell}_i \hat{\vec{F}} - \frac{\partial v_i}{\partial \tau} - \sum_{j=1}^{m} (\lambda_i - \lambda_j) \frac{\partial v_i}{\partial \zeta_j} , \qquad (1.54a)$$

$$i = 1,\ldots,m,$$

$$\frac{\partial V_{1j}}{\partial t} + \lambda_j \frac{\partial V_{1j}}{\partial x} = \vec{\ell}_j \hat{\vec{F}}, \qquad j = m+1,\ldots,N. \qquad (1.54b)$$

The right-hand side of (1.54a) is most of all a function of the variables $\xi_1,\ldots,\xi_m$, and consequently the V_{1i} are also functions of these same variables. Therefore, in order to avoid the secular increase of V_{1i} we require the right-hand side of (1.54a) to be zero. It leads to the m equations

$$\frac{\partial v_i}{\partial \tau} = \vec{\ell}_i \hat{\vec{F}} - \sum_{j=1}^{m} (\lambda_i - \lambda_j) \frac{\partial v_i}{\partial \zeta_j} , \quad i = 1,\ldots,m \qquad (1.55)$$

governing the functions $v_1,\ldots,v_m$. In this case the V_{1i} are also zero. The right-hand side of the system (1.54b) is completely determined and therefore V_{1j} may be found uniquely. As a matter of fact, the procedure for determining V_{1i} and V_{1j} is similar to that of the iterative method.

Thus, in the first approximation, the evolution equations following from (1.55) may be written in the form

$$\frac{\partial v_i}{\partial \tau} + \sum_{j=1}^{m} (\lambda_i - \lambda_j) \frac{\partial v_i}{\partial \zeta_j} = \vec{\ell}_i \hat{\vec{F}}\{v_1,\dots,v_m\}. \tag{1.56}$$

These equations, generally speaking, coincide with the equations of the first approximation (1.9) obtained by the iterative method, provided that the change of variables (1.50) is taken into account. The evolution equations of higher approximations are constructed in a similar way. It must be emphasized, however, that in contrast to the initial equations, the simplified equations must be solved in the multidimensional space $\xi_1,\dots,\xi_m,\tau$, which is not always convenient in practice.

1.4 The spectral method

Now let us consider the spectral method for constructing evolution equations that is especially convenient for wave processes in media with arbitrary (not only weak) dispersion. In addition this gives the possibility of checking the correctness of the single wave approach [15, 18, 35, 42]. Here we start from system (1.1) rewritten in a generalized form

$$I \frac{\partial \vec{U}}{\partial t} + \hat{L}\vec{U} = \varepsilon \hat{M}\vec{U}, \tag{1.57}$$

where $\hat{L}$ is a linear operator with respect to x (a matrix, dependent on $\partial/\partial x$) and $\hat{M}$ is a nonlinear operator. For the sake of definiteness we assume $\hat{L}$ to be a differential operator. The wave field may be represented as the sum of elementary linear solutions in the form of the sinusoidal progressive wavetrains

$$\vec{U} - \sum_m dk\, u_k^m \vec{r}_k^m \exp\{i(\omega_k^m t - kx)\} + c.c., \tag{1.58}$$

where $\vec{r}_k^m$ is the right eigenvector of the operator $\hat{L}$ corresponding to the eigenvalue ω_k^m

$$\hat{L}(-ik)\vec{r}_k^m = -i\omega_k^m \vec{r}_k^m, \tag{1.59a}$$

$$\mathrm{Det}|\hat{L}(-ik) + i\omega_k^m I| = 0 \tag{1.59b}$$

and u_k^m is the spectral amplitude of the wave. Here m denotes the number of the mode and c.c. is the complex conjugate. Because (1.58) is not the exact solution of (1.58) at $\varepsilon \neq 0$, we shall consider the expression (1.58) as a formal change of variables, assuming u_k^m in (1.58) to be time-dependent. Introducing (1.58) into the equation (1.5) and carrying out the inverse Fourier-transformation with respect to the space coordinate we arrive at an infinite system for determining the spectral amplitudes

$$\frac{du_k^m}{dt} = \varepsilon \sum_{m_1 m_2} \int dk_1 dk_2\, V_{k\,k_1 k_2}^{m\,m_1 m_2}\, u_{k_1}^{m_1} u_{k_2}^{m_2} \times$$

$$\times \exp\{i(\omega_{k_1}^{m_1} + \omega_{k_2}^{m_2} - \omega_k^m)t\}\delta(k_1 + k_2 - k), \tag{1.60}$$

where $V_{k\,k_2\,k_2}^{m\,m_1\,m_1}$ are found from the given operator $\hat{M}$ in an explicit form, and they depend only upon the wave numbers k and mode numbers m. Here we shall not discuss the technical details in the calculation of V in specific situations as this is not of great significance. According to the established terminology, V is called the matrix-coefficient. We must emphasize that system (1.60) is not an approximate one. On the contrary, it is an exact system equivalent in any sense to the initial system (1.57). It is obvious that not only the wave variables but also the external fields may be represented in the form of Fourier integrals. The same may be said about the parameters of inhomogeneity and nonstationarity. Therefore equations (1.60) do not govern only the interaction of waves, but also the interaction between waves and the outer field. Finally we point out that if the nonlinear operators in the right-hand side of (1.57) have higher terms (cubic terms for example) then additional higher (cubic) terms also appear in equations (1.60).

The system of equations (1.60) governing the spectral amplitudes is infinite. The solutions of this system in the general case are not known and therefore we must consider its simplification. It is clear that the strongest interaction of the spectral amplitudes will occur under conditions of synchronism when

$$k_1 + k_2 + k = 0, \quad \Delta \equiv \omega_{k_1}^{m_1} + \omega_{k_2}^{m_2} - \omega_k^m = 0 \tag{1.61}$$

or, at least, when $\Delta \sim \varepsilon$. If many modes are generated, then the conditions (1.61) are almost always, satisfied, and several types of waves interact. Particularly, if three waves are propagated in a nondispersive medium, then conditions (1.61) are always satisfied (see Figure 1.1). Therefore, single

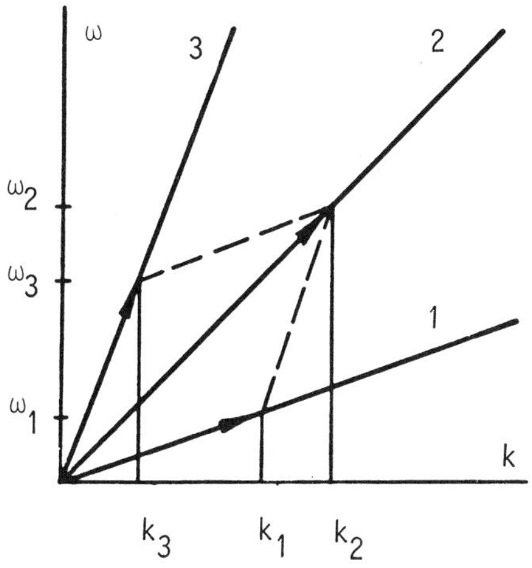

Figure 1.1. Diagram of ω-k plane in the case of three interacting waves of different type

waves can be considered separately provided the condition

$$V_{k\ k_1 k_2}^{m\ m_1 m_2} = 0, \quad m_{1,2} \neq m \tag{1.62}$$

is satisfied. Here the wave with the wave number m propagates independently. The single-wave approximation is also possible when (1.62) is not satisfied, but the boundary and initial conditions correspond to the generation of only one definite mode. Let us consider once more the situation depicted in Fig. 1.1 and let us limit ourselves to three interacting waves. In this case equations (1.60) may be rewritten as

$$\frac{da_1}{dt} = -Va_2a_3^*, \tag{1.63a}$$

$$\frac{da_2}{dt} = Va_1a_2, \tag{1.63b}$$

$$\frac{da_3}{dt} = -Va_1^*a_2, \tag{1.63c}$$

where a_i^* are the complex conjugate amplitudes. Here we have used the fact that in the case of conservative systems the matrix-coefficients all have the same value $V > 0$ [8]. If mainly the second mode is to be generated ($a_2 >> a_{1,3}$), then for the small corrections of $a_{1,3}$ we may use the equations

$$\frac{d^2a_{1,3}}{dt^2} = V^2a_2^2a_{1,3} \tag{1.64}$$

where a_2 is considered to have a constant value (the approximation of the given field). The exponential growth of $a_{1,3}$ is obvious, and hence the energy from the second mode will be "pumped" into the first and third modes. Therefore the single-wave approximation is not valid. If mainly the first mode (or the third mode) is to be generated, then the small corrections to $a_{2,3}$ in the approximation of the given field are governed by

$$\frac{d^2a_{2,3}}{dt^2} = -Va_1^2a_{2,3}. \tag{1.65}$$

Hence, the amplitudes a_2, a_3 do not increase in the course of time, and therefore the single-wave approximation is valid.

Thus the spectral method gives a rather convenient possibility of justifyi the single-wave approach. Such a checking process is very cumbersome when using the iterative and the asymptotic methods. It should be pointed out that this analysis is carried out purely on the basis of the dispersion relation, without detailed knowledge of the nonlinear interaction between the waves i.e. actually before solving the problem itself. If the single-wave approximation is valid then equations (1.60) are simplified and we arrive (the index m is neglected) at the equations

$$\frac{du_k}{dt} = \int dk_1 dk_2 V_{k\,k_1 k_2}\, u_{k_1} u_{k_2}\, \exp\{i\Delta t\}\delta(k_1 + k_2 - k). \tag{1.66}$$

Now applying the Fourier transform with respect to the space coordinate x, system (1.66) is transformed into a single equation, that is, generally speaking, an integro-differential equation.

Thus the scheme for constructing single-wave evolution equations according to the spectral method is the following:

- an exact transformation into the equations of the spectral amplitudes;

- the investigation of the conditions of synchronism, and the neglecting of terms responsible for the interaction (in the first approximation);

- the inverse Fourier transform of the simplified system governing the spectral amplitudes.

The next approximations are calculated through an iterative procedure. The first step is to calculate the small corrections for other modes (u_k^n, $n \neq m$) generated by the main wave with the mode number m. Here the equation

$$\frac{du_k^m}{dt} = \varepsilon \int dk_1 dk_2 V_{k\,k_1 k_2}^{n\,m\,m}\, u_{k_1}^m u_{k_2}^m\, \exp(\Delta t)\delta(k_1 + k_2 - k) \tag{1.67}$$

may be used that follows from the equation (1.60) according to the perturbation procedure. The next step is to introduce u_k^m into the expressions $u_{k_1}^n\ u_{k_2}^m$ in the equation (1.60). Then the cubic terms appear in the equation (1.66) and in fact we obtain the evolution equation of the second approximation etc.

The spectral method is mainly used for Hamiltonian systems [18]. The Hamiltonian functions are known for many waves in various physical problems. According to the so-called Hamiltonian formalism, the canonical field variables determined from the Hamiltonian functions are expanded into the series (1.58). The equations for the spectral amplitudes as well as the initial equations take the form of the Hamiltonian equations

$$\frac{du_k^m}{dt} = -i\,\frac{\delta H}{\delta u_k^{m*}}\,, \qquad \frac{du_k^{m*}}{dt} = i\,\frac{\delta H}{\delta u_k^m}\,, \tag{1.68}$$

where H is the Hamiltonian function, u_k^{m*} is the complex conjugate of u_k^m and $\delta H/\delta u$ is the variational derivative. It is essential that the equations (1.68) coincide exactly with the equations (1.60). The Hamiltonian formalism is effective in many cases. The reader can turn for details to [121].

Application of the spectral method is also effective with Lagrangian systems [1, 91]. Instead of the Lagrangian $L(\vec{U},\vec{U}_t,\vec{U}_x)$ it is more convenient to introduce the averaged Lagrangian

$$\mathcal{L} = \int L d\Theta_1 d\Theta_2 \ldots, \tag{1.69}$$

where $\Theta_i = \omega_i t - k_i x$ are the phases of the interacting waves. As far as the averaged Lagrangian is a function of $\omega = \partial\Theta/\partial t$ and of sums like $\Theta_1 + \Theta_2 - \Theta$ satisfying the conditions of synchronism over both ω and k, the calculus of variation yields [55, 116]

$$\frac{d}{dt}\frac{\partial \mathcal{L}}{\partial \omega_k} - \frac{\partial \mathcal{L}}{\partial \Theta_k} = 0 \tag{1.70}$$

which are equivalent to (1.60). The variational principle described here is fully proved elsewhere [116].

The spectral approach is widely used in numerical schemes of integration of the evolution equations [94]. In this case, as a rule, periodicity (over x) is taken into account. The spectrum of wave numbers is discrete and the system of equations of types (1.68) and (1.70) is transformed to a system of ordinary differential equations. The order of this system may be high, but the problem is still solvable using contemporary computers. It should be emphasized that the structure of the dispersion relations is not very important when the equations (1.68) are investigated. It is reflected only in the structure of the algebraic functions V(k). On the other hand, the order of the evolution equation depends directly on the structure of the dispersion relation.

Finally we note that the spectral approach may be applied either to the initial system or to the evolution equation. The latter is justified when the analytical solution of the evolution equation is not known.

1.5 Strongly nonlinear systems

There are no regular methods for constructing the evolution equations governing waves in media with the arbitrary non-linearity. Nevertheless, several approaches exist that are able to simplify the initial system. Here we shall demonstrate the effectiveness of the transformation to the Riemann invariants in case of hyperbolic systems. The physical example considered describes acoustical waves in an isentropic weakly inhomogeneous gas.

The initial system of equations is taken in the form

$$\frac{\partial u}{\partial t} + u\,\frac{\partial u}{\partial x} + \frac{1}{\partial}\,\frac{\partial p}{\partial x} = \varepsilon f(x), \tag{1.71a}$$

$$\frac{\partial \rho}{\partial t} + \frac{\partial}{\partial x}\,(\rho u) = 0, \tag{1.71b}$$

$$p = \text{const}\cdot\rho^{\gamma}. \tag{1.71c}$$

Here u is the particle velocity, ρ - the density, p - the pressure, γ - the ratio of specific heats and $\varepsilon f(x)$ is the external field describing the weak inhomogeneity of all physical parameters. We shall introduce the Riemann invariants analogously to the case of a homogeneous medium

$$J_{\pm} = u \pm \frac{c - c_0(\chi)}{\gamma - 1}\ , \tag{1.72}$$

where $c_0(\chi)$ is the sound velocity in the unperturbed medium which depends on the space coordinate. Treating (1.72) as a formal change of variables we rewrite the system (1.71) in terms of $J_{\pm}$

$$\frac{\partial J_\pm}{\partial t} + [\pm c + \alpha J_\pm + \beta J_\mp] \frac{\partial J_\pm}{\partial x} =$$

$$= - \frac{2\varepsilon}{\gamma - 1} [\alpha J_\pm + \beta J_\mp] \frac{dc}{d\chi} , \qquad (1.73a)$$

$$\alpha = \frac{1}{4} (\gamma + 1); \; \beta = \frac{1}{4} (3 - \gamma). \qquad (1.73b)$$

The system (1.73) is strongly nonlinear and the equations for J_+ and J_- are coupled. Only when $\beta = 0$, i.e. $\gamma = 3$ do the equations uncouple, and in spite of the strong nonlinearity an exact solution can then be found. This is actually a singular case and we shall not restrict ourselves to it. Further we assume $\beta \neq 0$ and consider the following situation. At the boundary $x = 0$ let a wave propagating so that $x > 0$ be generated. Should the gas be homogeneous, system (1.73) yields

$$J_+ \neq 0, \quad J_- = 0. \qquad (1.74)$$

In a weakly inhomogeneous medium we have $J_- = \varepsilon$. Hence equation (1.73a) does not contain the small parameter only on its right-hand side, but also on its left-hand side the small parameter exists as a multiplier of the terms responsible for the interaction ($\beta J_\mp \partial J_\pm / \partial x$). According to the regular iterative method in the first approximation the equation for J_- is linearized to give

$$\frac{\partial J_-}{\partial t} + (-c + \beta J_+) \frac{\partial J_-}{\partial x} = \frac{2\varepsilon\beta}{\gamma - 1} J_+ \frac{dc}{d\chi} . \qquad (1.75)$$

This equation may be solved by the method of characteristics and its solution may be represented in the functional form $J_- = \varepsilon\hat{\phi}(J_+, \chi)$. Knowing the variable J_-, the equation for J_+ is obtainable in closed form and the corresponding evolution equation then reads

$$\frac{\partial J_+}{\partial t} + (c + \alpha J_+) \frac{\partial J_+}{\partial x} = - \frac{2\varepsilon\alpha}{\gamma - 1} J_+ \frac{dc}{d\chi} - \varepsilon\beta\hat{\phi} \frac{\partial J_+}{\partial x} . \qquad (1.76)$$

In principle, this iterative procedure may be continued indefinitely but the calculations are rather awkward due to the strong nonlinearity.

There is, however, an important physical restriction that must be taken into account when applying the simplification methods. The strong nonlinearity leads rapidly to the "gradient catastrophe" (shock wave formation). It might thus happen that the weak effects of the small order described by terms on the right-hand side are not able to develop and to play a significant role in the production of distortion. Consequently these effects (weak inhomogeneity, weak dispersion and dissipation) may be taken into account according to the usual perturbation theory. The equations of type (1.76) are again not applicable for strong shock waves when the motion is not isentropic and also reflections from the shock-wave front are not taken into account [53, 54]. All this essentially restricts the simplification of strongly nonlinear systems.

1.6 The comparative estimation of accuracy

As shown above, the evolution equations of the first approximation deduced by various methods coincide. The higher approximations again differ, which shows explicitly the difference in terms of convergence between the various evolution equations. Strict theorems for the validity and uniqueness of solutions similar to those in the perturbation theory of nonlinear oscillations [111] is absent up to now. Here we present a comparative analysis treating the model nonlinear wave equation

$$\frac{\partial^2 u}{\partial t^2} - \frac{\partial^2 u}{\partial x^2} = \varepsilon \frac{\partial^2 u^2}{\partial x^2} . \qquad (1.77)$$

It is equivalent to the matrix system (1.1) with

$$\vec{U} = \begin{vmatrix} u \\ v \end{vmatrix}, \quad A = \begin{vmatrix} 0 & 1 \\ 1 & 0 \end{vmatrix}, \quad \hat{\vec{F}} = \begin{vmatrix} 0 \\ -\frac{\partial u^2}{\partial x} \end{vmatrix} . \qquad (1.78)$$

In terms of the usual equations it reads

$$\frac{\partial u}{\partial t} + \frac{\partial v}{\partial x} = 0, \qquad (1.79a)$$

$$\frac{\partial v}{\partial t} + \frac{\partial u}{\partial x} = - \varepsilon \frac{\partial u^2}{\partial x} . \qquad (1.79b)$$

Equation (1.77), or the system (1.79), permits the construction of the Riemann solution. Assuming $v = v(u)$, the governing system (1.79) may be represented as

$$\frac{\partial u}{\partial t} + \frac{dv}{du}\frac{\partial u}{\partial x} = 0, \tag{1.80a}$$

$$\frac{dv}{du}\frac{\partial u}{\partial t} + (1 + 2\varepsilon u)\frac{\partial u}{\partial x} = 0. \tag{1.80b}$$

A nontrivial solution of the system (1.80) exists provided the determinant of the system vanishes.

This condition yields

$$\left(\frac{dv}{du}\right)^2 = 1 + 2\varepsilon k, \tag{1.81a}$$

$$v = \pm \frac{(1 + 2\varepsilon u)^{3/2} - 1}{3\varepsilon}. \tag{1.81b}$$

Now it is easy to find the exact solution to (1.79). For a wave propagatin in the $x > 0$ direction we obtain

$$u = F(x - V(u)t), \tag{1.82a}$$

$$V(u) = (1 + 2\ u)^{1/2}, \tag{1.82b}$$

where F is an arbitrary function determined by the initial (boundary) conditions. The existence of the exact solution (1.82) makes the accuracy estimate bounded.

The iterative method. We shall represent now the governing equation (1.77) in the normal form. Introducing

$$u = \frac{1}{2}(w + w_-); \quad v = \frac{1}{2}(w - w_-), \tag{1.83}$$

the governing system takes the form

$$\frac{\partial w}{\partial t} + \frac{\partial w}{\partial x} = - \frac{\varepsilon}{4} \frac{\partial}{\partial x} (w + w_-)^2, \tag{1.84a}$$

$$\frac{\partial w_-}{\partial t} + \frac{\partial w_-}{\partial x} = \frac{\varepsilon}{4} \frac{\partial}{\partial x} (w + w_-)^2. \tag{1.84b}$$

According to the iterative procedure (see Section 1.1) we get in the zeroth approximation the system (1.84) with $\varepsilon = 0$ and

$$w = w(x - t), \quad w_- = 0, \tag{1.85}$$

which is equivalent to

$$u = u(x - t), \quad v = u. \tag{1.86}$$

This coincides with the exact solution when $\varepsilon = 0$. The first approximation is governed by

$$\frac{\partial w}{\partial t} + \frac{\partial w}{\partial x} = - \frac{\varepsilon}{4} \frac{\partial w^2}{\partial x}, \tag{1.87}$$

$$\frac{\partial w_-}{\partial t} - \frac{\partial w_-}{\partial x} = \frac{\varepsilon}{4} \frac{\partial w^2}{\partial x}. \tag{1.88}$$

On account of the hyperbolicity of the initial system the evolution equation (1.87) has the typical form of the Riemann wave equation with the velocity

$$V_1 = 1 + \frac{\varepsilon}{2} w. \tag{1.89}$$

Noting that $w \sim 2u + O(\varepsilon)$, the final result is

$$V_1 \simeq 1 + \varepsilon u. \tag{1.90}$$

This result is easily obtained from the exact solution (1.82) by using the series representation and retaining only two terms. In order to construct the evolution equation of the second approximation we need w_- to within an error ε. Since w depends on x - t, the variable w_- must also depend on the same combination. After integration of (1.88) w_- is determined by

$$w_- = -\frac{\varepsilon}{8} w^2. \tag{1.91}$$

Substituting this result into (1.84a) the required evolution equation takes the form

$$\frac{\partial w}{\partial t} + \frac{\partial w}{\partial x} = -\frac{\varepsilon}{2} w \frac{\partial w}{\partial x} + \frac{3\varepsilon^2}{16} w^2 \frac{\partial w}{\partial x}. \tag{1.92}$$

This equation is again hyperbolic and the velocity V_2 is easily determined as

$$V_2 = 1 + \frac{3}{2} w - \frac{3\varepsilon^2}{16} w^2. \tag{1.93}$$

Now, the relations (1.83) yield

$$u \simeq \frac{w}{2} - \frac{\varepsilon}{16} w^2, \tag{1.94}$$

which is equivalent to

$$w \simeq 2u + \frac{\varepsilon}{2} u^2. \tag{1.95}$$

In terms of u the velocity V_2 may be calculated as

$$V_2 = 1 + \varepsilon u - \frac{1}{2} \varepsilon u^2. \tag{1.96}$$

Again, this result is easily obtained from the exact solution (1.82) by retaining three terms in the series. It is easily understood that the exactness of the solution obtained by the iterative method increases with higher approximations. The evolution equation of the n-th approximation is the representation of the initial equation for the Riemann wave with an error of order ε^n. The exact hyperbolic equations are correct in a bounded time interval up to t_* - the shock wave formation time. The evolution equations are correct for $t < t_n$ with the error estimate

$$\max |t_n - t_*| < O(\varepsilon^n). \qquad (1.97)$$

The asymptotic scheme. Introducing the independent variables

$$\xi = x - t, \ \tau = \varepsilon t, \qquad (1.98)$$

the governing system may be rewritten in the form

$$-\frac{\partial u}{\partial \xi} + \frac{\partial v}{\partial \xi} = -\varepsilon \frac{\partial u}{\partial \tau}, \qquad (1.99a)$$

$$\frac{\partial u}{\partial \xi} - \frac{\partial v}{\partial \xi} = -\varepsilon \frac{\partial v}{\partial \tau} - \varepsilon \frac{\partial u^2}{\partial \xi}. \qquad (1.99b)$$

The unknown functions u and v are represented in series form by

$$u,v = (u_0, v_0) + \varepsilon(u_1, v_1) + \varepsilon^2(u_2, v_2) + \dots . \qquad (1.100)$$

In the zeroth approximation system (1.99) yields

$$-\frac{\partial u_0}{\partial \xi} + \frac{\partial v_0}{\partial \xi} = 0, \qquad (1.101a)$$

$$\frac{\partial u_0}{\partial \xi} - \frac{\partial v_0}{\partial \xi} = 0, \qquad (1.101b)$$

with the trivial result

$$v_0 = u_0(\xi, \tau). \qquad (1.102)$$

The first approximation is governed by

$$-\frac{\partial u_1}{\partial \xi} + \frac{\partial v_1}{\partial \xi} = -\frac{\partial u_0}{\partial \tau}, \qquad (1.103a)$$

$$-\frac{\partial v_1}{\partial \xi} + \frac{\partial u_1}{\partial \xi} = -\frac{\partial v_0}{\partial \tau} - \frac{\partial u_0^2}{\partial \xi}. \qquad (1.103b)$$

Since the matrix A - I is degenerate, nontrivial solutions exist provided the orthogonality condition is satisfied. This condition (see (1.29)) gives the required evolution equation of the first approximation

$$\frac{\partial u_0}{\partial \tau} + u_0 \frac{\partial u_0}{\partial \xi} = 0, \tag{1.104}$$

which coincides with (1.87) with an error of $O(\varepsilon)$. Using (1.104) the solution of the system (1.103) is represented by

$$u_1 = u_+(\xi,\tau), \tag{1.105a}$$

$$v_1 = u_+(\xi,\tau) + \frac{1}{2} u_0^2, \tag{1.105b}$$

where u_+ must be determined from the evolution equation of the second approximation. This equation must be constructed from the system

$$-\frac{\partial u_2}{\partial \xi} + \frac{\partial v_2}{\partial \xi} = -\frac{\partial u_1}{\partial \xi}, \tag{1.106a}$$

$$-\frac{\partial v_2}{\partial \xi} + \frac{\partial u_2}{\partial \xi} = -\frac{\partial v_1}{\partial \tau} - 2\frac{\partial}{\partial \xi}(u_0 u_1). \tag{1.106b}$$

Once more, the orthogonality condition gives the evolution equation of the second approximation

$$\frac{\partial u_+}{\partial \tau} + \frac{\partial}{\partial \xi}(u_0 u_1) - \frac{u_0^2}{2}\frac{\partial u_0}{\partial \xi} = 0. \tag{1.107}$$

Its solution is

$$u_+ = \frac{1}{2} u_0^2, \tag{1.108}$$

and the second-order corrections, consequently, are

$$u_1 = \frac{1}{2} u_0^2, \quad v_2 = u_0^2. \tag{1.109}$$

Now it can easily be concluded that the asymptotic method gives small corrections to the main solution u_0 which once determined, do not then change. Due to the hyperbolicity, the solution of (1.104) is bounded up to time t_1. Hence the existence of the series representation is valid up to the same time. Therefore the estimate

$$\max |t_n - t_*| < O(\varepsilon) \tag{1.110}$$

determines the exactness of the evolution equations obtained by the asymptotic method.

The spectral method. We seek the solution to equation (1.77) in the form of Fourier integrals which describe waves propagating in opposite directions

$$u = \int dk \{A_k(t) \exp[ik(x - t)] + B_k(t)\exp[ik(x + t)] + + \text{c.c.}\}. \tag{1.111}$$

Substituting (1.111) into the equation (1.77) we obtain exact equations for the spectral amplitudes $A_k(t)$, $B_k(t)$. Here we represent the equation governing $A_k(t)$

$$\begin{aligned}\frac{dA_k}{dt} = -\frac{i\varepsilon}{2k}\int dk_1 dk_2 \{&\delta(k_1 + k_2 - k)(k_1 + k_2)^2 [A_{k_1}A_{k_2} + \\ &+ B_{k_1}B_{k_2} \exp i(k_1 + k_2 + k)t + A_{k_1}B_{k_2} \exp(-k_1 + k_2 + k)t + \\ &+ A_{k_1}B_{k_2} \exp i(k_1 - k_2 + k)t] + \\ &+ \delta(k_1 - k_2 - k)(k_1 - k_2)^2 [A_{k_1}A^*_{k_2} \exp i(k_2 - k_1 + k)t + \\ &+ B_{k_1}B^*_{k_2} \exp i(k_1 - k_2 + k)t + A_{k_1}B^*_{k_2} \exp i(k - k_1 - k_2)t] + \\ &+ \text{c.c.}\}.\end{aligned} \tag{1.112}$$

Here A^*_k, B^*_k are the complex conjugate amplitudes. Taking into account the properties of δ-functions we see that only the first terms in braces have a resonance character. Therefore in the first approximation all terms containing the multiplier exp $i\Delta_\pm t$, $\Delta_\pm = k_1 \pm k_2 \mp k$ may be neglected. Thus we find

$$\frac{dA_k}{dt} = - \frac{i\varepsilon k}{2} \int dk_1 dk_2 \, \{A_{k_1} A_{k_2} \, \delta(k_1 + k_2 - k) +$$

$$+ A_{k_1} A_{k_2} \, \delta(k_1 - k_2 - k) + c.c.\}. \tag{1.113}$$

This equation does not contain the amplitude of "another" wave B_k, and is closed with respect to A_k. Reconstructing the wave field by the expression

$$v = \int dk \, A_k(t)[\exp ik(x - t)] \tag{1.114}$$

we obtain the equation governing v in the single-wave approximation

$$\frac{\partial v}{\partial t} + \frac{\partial v}{\partial x} + \frac{\varepsilon}{2} \frac{\partial v^2}{\partial x} = 0. \tag{1.115}$$

It is clear that this equation was obtained earlier by making use of iterative and asymptotic methods. However, this reconstruction with the aid of the spectral method has explicitly demonstrated that in the first approximation the interaction between waves travelling in opposite directions is absent. This is not an obvious result when the other methods are used.

Discussion. Comparing the estimates (1.97) and (1.110) it is easily concluded that the iterative method has the better convergence to the exact solution. The spectral method, as seen from above discussions actually uses the iterative approach to construct the evolution equations, and therefore there is no need to treat it independently. The same is true with respect to the modified asymptotic method.

It is apparent that these estimates are obtained for the model example (1.77), and the need for more precise estimates and strict proofs is obvious. We should like to emphasize, however, that the correctness of the evolution equations of the first approximation obtained by the methods discussed above is the same.

2 The quasiplane waves in nonlinear media and the evolution equations

In practice the limited size of a transducer and/or the inhomogeneity of the medium parameters lead to a complicated field structure that no longer depends on one space coordinate (say, upon a Cartesian coordinate). In this case the methods described in Chapter 1 are not applicable. If, however, the field variables change slowly in the transverse direction, or the wave path is slightly curved in the course of propagation, then the notion of quasiplane waves may be used. The slow changes are meant here to be with respect to the wavelength as to the main parameter of the dimension for the excitation. Using the relation of the wavelength to the characteristic scale parameters of the wavefield as a small parameter, it is possible to construct certain asymptotic methods for describing such a wave process. Such asymptotic methods are well known in the linear theory for waves in media with slowly changing parameters. Usually they are called geometrical methods (the method of geometrical optics, the method of geometrical acoustics). A similar approach may also be derived for nonlinear waves, but the mathematical side of the problem is usually not discussed in detail. In this Chapter an attempt is made to present the geometrical method for nonlinear waves and to analyze both the mathematical and the physical aspects of the problem. Therefore, beside the strict algorithms, the physical approaches are given as examples.

2.1 The asymptotic scheme for propagation along the linear rays

The initial system describing the quasiplane waves is the following

$$I \frac{\partial \vec{U}}{\partial t} + \sum_{i=1}^{3} A_i \frac{\partial \vec{U}}{\partial x_i} = \varepsilon \hat{\vec{F}} \{\vec{u}, \vec{x}, t, \vec{X}, \tau\}, \tag{2.1a}$$

$$\vec{X} = \varepsilon \vec{x}, \quad \tau = \varepsilon t, \quad \varepsilon \ll 1, \tag{2.1b}$$

where $\vec{x}\{x_1, x_2, x_3\}$ is the space vector, x_i are the space coordinates, $A_i = \{A_1, A_2, A_3\}$ are $n \times n$ matrices with the components independent of field variables. We shall assume that when $\varepsilon = 0$ the eigenvalues of the matrices A_i are real and distinct, i.e. the system (2.1) is hyperbolic. Hence solutions of type (1.2) exist along certain curves in the space. These curves are actually the wave paths for the main waves. Further we limit ourselves to the single-wave approximation when the number of main waves is one. We assume also $A_i = A_i(\vec{X})$ which corresponds to the usual wave propagation situation in smoothly inhomogeneous media.

We seek a solution to (2.1) similar to (1.23) in the form of an asymptotic series

$$\vec{U} = \vec{U}_0 + \varepsilon\vec{U}_1 + \varepsilon^2\vec{U}_2 + \dots . \tag{2.2}$$

Every term in this series is a function of the following variables

$$\xi = \varphi(\vec{r}) - t, \quad \vec{X} = \varepsilon\vec{r}, \tag{2.3}$$

where according to the established terminology φ is called eikonal. This must be determined by the further analysis.

Substituting (2.2) into the system (2.1) and taking the expressions (2.3) into account we obtain from the powers $\varepsilon^0, \varepsilon^1, \varepsilon^2, \dots, \varepsilon^n$

$$\left(\sum_{i=1}^{3} A_i \frac{\partial\varphi}{\partial x_i} - I\right) \frac{\partial\vec{U}_0}{\partial\xi} = 0, \tag{2.4}$$

$$\left(\sum_{i=1}^{3} A_i \frac{\partial\varphi}{\partial x_i} - I\right) \frac{\partial\vec{U}_1}{\partial\xi} = \hat{\vec{F}}\left\{U_0, \varepsilon = 0\right\} - \sum_{i=1}^{3} A_i \frac{\partial\vec{U}_0}{\partial X_i}, \tag{2.5}$$

$$\left(\sum_{i=1}^{3} A_i \frac{\partial\varphi}{\partial x_i} - I\right) \frac{\partial\vec{U}_2}{\partial\xi} = \frac{\delta\vec{F}}{\delta U_0} U_1 - \frac{\partial\vec{F}}{\partial\varepsilon} - \sum_{i=1}^{3} A_i \frac{\partial\vec{U}_1}{\partial X_i}, \tag{2.6}$$

$$\dots$$

$$\left(\sum_{i=1}^{n} A_i \frac{\partial\varphi}{\partial x_i} - I\right) \frac{\partial\vec{U}_n}{\partial\xi} = \hat{\vec{H}}\left\{\vec{U}_0, \dots, \vec{U}_{n-1}\right\} - \sum_{i=1}^{3} A_i \frac{\partial\vec{U}_{n-1}}{\partial X_i}, \tag{2.7}$$

respectively. Exactly as in the one-dimensional case (see 1.2), every successive approximation raises the accuracy when determining the field variables at a distance (ε^{-1}). The solution of the zeroth approximation, similar to (1.27) is

$$\vec{U}_0 = v_0(\xi,\vec{\chi})\vec{r}, \tag{2.8}$$

provided the condition

$$M = \mathrm{Det}\left|\sum_{i=1}^{3} A_i \frac{\partial\varphi}{\partial x_i} - I\right| = 0 \tag{2.9}$$

is satisfied. Here v_0 is an arbitrary scalar function that must be determined subject to the initial conditions and $\vec{r}$ is the right eigenvector of the matrix $\sum_{i=1}^{3} A_i(\partial\varphi/\partial x_i)$. The condition (2.9) is used for determining the eikonal and we shall discuss it in detail. In the one-dimensional case obviously $\partial\varphi/\partial x$ coincides with $\lambda^{-1}(x)$, i.e. we have $\varphi = \int dx/\lambda$. In the general case the system (2.9) governing the eikonals represents the Hamilton-Jacobi system of the ordinary differential equations of the first order. The mathematical methods for solving this system are well-known [52]. Since $M = M(\vec{r},\nabla\varphi) = 0$ the characteristic system for every eikonal $\varphi(\vec{r})$ has the form

$$\frac{dx_i}{\partial M/\partial p_i} = -\frac{dp_i}{\partial M/\partial x_i} = \frac{d\varphi}{\sum_{i=1}^{3} p_i(\partial M/\partial p_i)} . \tag{2.10}$$

Here $p_i = \partial\varphi/\partial x_i$. The ray equations in the parametric form now follow easily as

$$\frac{dx_i}{d\tau} = \frac{\partial M}{\partial p_i}, \quad \frac{dp_i}{d\tau} = -\frac{\partial M}{\partial x_i}, \quad \frac{d\varphi}{d\tau} = \sum_{i=1}^{3} p_i \frac{\partial M}{\partial p_i} . \tag{2.11}$$

This system determines the rays on the basis of the zeroth-order equations (2.4), and these rays do not depend upon the nonlinearity involved into the system. Let us denote the velocity along these rays by λ. Then the differential $ds = \lambda d\tau$ is actually a line element of the ray governed by the equations (2.11). If the system of the equations (2.1) is of the second order describing the wave propagation in an isotropic medium, then the condition M = 0 yields the known eikonal equation

$$(\nabla\varphi)^2 = \lambda^{-2}. \tag{2.12}$$

Here λ is the eigenvalue of the scalar matrix $\vec{A}\cdot\vec{n}$, where $\vec{n}$ denotes the unit vector normal to the eikonal and $\vec{A}$ - the vector matrix with the components A_i.

The arbitrary functions $v_0(\xi,\vec{\chi})$ introduced by expression (2.8) exist provided the orthogonality conditions (1.28) for the equation (2.5) are satisfied. Consequently, we require

$$\vec{\ell}A_i \frac{\partial}{\partial\chi_i}\vec{U}_0 = \vec{\ell}\hat{\vec{F}}, \tag{2.13}$$

where $\vec{\ell}$ is the left eigenvector to the matrix $\vec{A}\cdot\vec{n}$. Since $A_i\,\partial/\partial\chi_i = \vec{A}\,\vec{n}\partial/\partial s$, the condition (2.13) may be transformed into

$$\vec{\ell}\lambda\,\frac{\partial\vec{U}_0}{\partial s} = \vec{\ell}\hat{\vec{F}}. \tag{2.14}$$

Here the difference from the one-dimensional case is in the ray coordinate s. Introducing the function v_0 into the condition (2.14) we obtain

$$\frac{\partial v_0}{\partial s} = \frac{1}{\lambda}\,\frac{\vec{\ell}\cdot\vec{F}}{\vec{\ell}\cdot\vec{r}} = \hat{f}_0(v_0). \tag{2.15}$$

To the first order, equation (2.6) yields

$$\frac{\partial v_1}{\partial s} = \frac{\vec{\ell}}{\lambda(\vec{\ell}\cdot\vec{r})}\left\{\frac{\delta\hat{\vec{F}}}{\delta\vec{U}_0}v_1\vec{r} + \frac{\partial\hat{\vec{F}}}{\partial\varepsilon}\right\} = \hat{f}_1(v_0,v_1). \tag{2.16}$$

The equations for the higher order approximations are derived in a similar way. To the n-th order the evolution equation has the form

$$\frac{\partial v_n}{\partial s} = \frac{\vec{\ell}}{\lambda(\vec{\ell}\cdot\vec{r})}\,\hat{\vec{H}}\left\{v_0\vec{r},\ldots,v_n\vec{r}\right\} = \hat{f}_n(v_0,\ldots,v_n). \tag{2.17}$$

Thus the wave field is determined in all approximations by equations (2.15)-

(2.17). The field variables depend on the ray coordinate s which is determined by the variable coefficients of the initial system of the equations. Nonlinear effects develop in the course of propagating along the "linear" rays.

Example 2.1. Let us consider the simplest example in acoustics which illustrates the joint influence of two factors: the nonlinearity and the inhomogeneity [90]. The initial equations are written in the form

$$\frac{\partial \vec{u}}{\partial t} + (\vec{u}\nabla)\vec{u} + \frac{\nabla p}{\rho} = \vec{f}, \tag{2.18a}$$

$$\frac{\partial \rho}{\partial t} + \operatorname{div} \rho\vec{u} = 0, \tag{2.18b}$$

$$\frac{\partial p}{\partial t} + (\vec{u}\nabla p) + \gamma p \operatorname{div} \vec{u} = 0. \tag{2.18c}$$

Here $\vec{u}$ is the particle velocity, ρ and p are the density and the pressure, respectively, γ is the adiabatic exponent and $\vec{f}$ is the "external force" (i.e. the acceleration due to gravity). The latter is responsible for the inhomogeneity of the medium, determining the hydrostatic equilibrium

$$\nabla\bar{p} = \bar{\rho}\,\vec{f}, \tag{2.19}$$

where $\bar{p}$ and $\bar{\rho}$ are the static distributions of pressure and density, respectively. Let us introduce the vector

$$\vec{U} = \begin{vmatrix} \vec{u} \\ \rho \\ p \end{vmatrix} \tag{2.20}$$

in order to represent the system (2.18) in matrix form. Here $\vec{u}$ includes the three components of the particle velocity. Next we introduce the unit vector $\vec{n}$ along the direction of wave propagation (the vector $\vec{n}$ is the sum of basis vectors). Now the matrix A and the functional $\hat{\vec{F}}$ (see the equation (2.1)) are represented by

$$A = \begin{vmatrix} 0 & 0 & \vec{n}/\rho \\ \vec{n}\bar{\rho} & 0 & 0 \\ \vec{n}\gamma\bar{p} & 0 & 0 \end{vmatrix}, \tag{2.21a}$$

$$\hat{\vec{F}} = \begin{vmatrix} \vec{f} - (\vec{u}\nabla)\vec{u} + (1/\rho - 1/\bar{\rho})\nabla p \\ \bar{\rho} \operatorname{div} \vec{u} - \operatorname{div} \rho\vec{u} \\ -\vec{u}\nabla p - \gamma(p - \bar{p}) \operatorname{div} \vec{u} \end{vmatrix}. \tag{2.21b}$$

The next step is to introduce the eikonal φ and the new variables (2.3). In the zeroth approximation we obtain the matrix $M = |\vec{A}\nabla\varphi - I|$ which after introducing the matrix (2.21a) has the form

$$M = \begin{vmatrix} -1 & 0 & \vec{n}\nabla\varphi/\bar{\rho} \\ \vec{n}\bar{\rho}\nabla\varphi & -1 & 0 \\ \vec{n}\gamma\bar{p}\nabla\varphi & 0 & -1 \end{vmatrix}. \tag{2.21c}$$

The eikonal equation is represented by $(\nabla\varphi)^2 = c^{-2}$ where $c^2 = \gamma\bar{p}/\bar{\rho}$ determines the unperturbed value of the sound velocity. The solution in the zeroth approximation is sought using the right eigenvector to the matrix $\vec{A}\nabla\varphi$

$$\vec{r} = \begin{Bmatrix} \vec{n}\nabla\varphi/\bar{\rho} \\ c^{-2} \\ 1 \end{Bmatrix}. \tag{2.22}$$

Here for definiteness we have used $v_0 = p$. Assuming $\vec{U} = p\vec{r}$ we have the obvious expressions $\vec{u} = \nabla\varphi p/\bar{\rho}$ and $\rho = p/c^2$. The condition for nonsecularity of the first order approximation gives the equation which determines v_0. This condition is actually the orthogonality condition where we also need the left eigenvector $\vec{\ell}$ to the matrix $\vec{A}\nabla\varphi$. It has the components

$$\vec{\ell} = \{\bar{\rho}c^2(\vec{n}\nabla)\varphi, c^2, 1\}. \tag{2.23}$$

Now, taking this expression into account, equation (2.14) yields

$$\frac{\partial p}{\partial s} + \frac{\gamma + 1}{2\bar{\rho}c^3} p \frac{\partial p}{\partial \xi} + \frac{\rho c^2}{2} \operatorname{div} \left(\frac{\nabla\varphi}{\bar{\rho}}\right) p = 0, \tag{2.24}$$

which is the required evolution equation. Thus, in this problem the functional $\hat{f}_0(p)$ includes two summands responsible for the nonlinearity and the inhomogeneity (the second and the third terms in (2.24), respectively). The last term in equation (2.24) contains the second derivative from the eikonal. According to [4, 116] φ may be easily determined by the ray characteristics

$$\Delta\varphi = \frac{1}{S} \frac{d}{ds} \left(\frac{S}{c}\right). \tag{2.25}$$

Here S is the cross-section of the ray tube. The final evolution equation for the "main" term of the asymptotic expansion has the following form

$$\frac{\partial p}{\partial s} + \frac{\alpha}{\bar{\rho}c^3} p \frac{dp}{d\xi} + \frac{c\bar{\rho}}{2S} \frac{d}{ds} \left(\frac{S}{c\bar{\rho}}\right) p = 0, \tag{2.26}$$

where $\alpha = (\gamma + 1)/2$.

2.2 The asymptotic scheme for propagation along nonlinear rays

The asymptotic scheme for propagation along linear rays described in the previous Section may be used for the calculation of field variables up to a certain time (distance) at which the ray trajectories are essentially distorted due to the nonlinear effects. In nonlinear acoustics such a distance is estimated, for example, in the case of waves with rectangular and triangular profiles [86, 90] and the corresponding distortion of ray trajectories is analysed in detail. We shall present this estimation in Chapter 4 in which we shall also consider a problem of nonlinear acoustics. There, in order to describe the nonlinear distortion of the ray trajectories, besides the expansion of field variables (2.2) we shall also use the expansion of the ray characteristics $\nabla\varphi$ [88, 89, 102-104]. Since the main nonlinear equation (2.15) contains the terms of the second order, the expansion of the eikonal gradient needs only the additional first order term

$$\nabla\varphi = \vec{\psi}_0 + \varepsilon\vec{\psi}_1(\vec{x}). \tag{2.27}$$

In contrast to the linear ray approximation, the correction $\vec{\psi}_1$ added here takes the slow change of the eikonal into account that is caused by the slow distortion of the wave profile. Now both the ray trajectory and the ray tube parameters depend on the wave field. The eikonal gradient determined with an accuracy of $O(\varepsilon)$ makes possible the construction of the rays and the calculation with the same accuracy of the nonlinear change of the cross-section of the ray-tube. The equations governing the eikonal dependence on the cross-section of the ray-tube are based on (2.25) and may be presented in the form

$$\nabla(\vec{n}/S) = 0, \tag{2.28}$$

$$\vec{n} = \nabla\varphi / |\nabla\varphi|. \tag{2.29}$$

As far as $\nabla\varphi$ is determined by the unknown function $\vec{\psi}_1$, the system of the equations (2.28), (2.29) is not closed. Thus, in order to complete the system (2.28), (2.29), the function $\vec{\psi}_1$ must be known in the course of propagation along the ray trajectory.

We shall make use of the asymptotic series (2.2) and the expansion (2.27). Equations (2.1) yield (in powers of $\varepsilon^0, \varepsilon^1, \varepsilon^2, \dots, \varepsilon^n$)

$$(\vec{\vec{A}}\psi_0 - I)\frac{\partial\vec{U}_0}{\partial\xi} = 0, \tag{2.30}$$

$$(\vec{\vec{A}}\psi_0 - I)\frac{\partial\vec{U}_1}{\partial\xi} = \hat{\vec{F}}\{\vec{U}_0,\ \varepsilon = 0\} - \vec{A}\nabla\vec{U}_0 - \vec{\vec{A}}\psi_1\frac{\partial\vec{U}_0}{\partial\xi}, \tag{2.31}$$

$$(\vec{\vec{A}}\psi_0 - I)\frac{\partial\vec{U}_2}{\partial\xi} = \frac{\delta\hat{\vec{F}}}{\delta\vec{U}_0}\vec{U}_1 + \frac{\partial\hat{\vec{F}}}{\partial\varepsilon} - \vec{A}\nabla\vec{U}_1 - \vec{\vec{A}}\psi_1\frac{\partial\vec{U}_1}{\partial\xi}, \tag{2.32}$$

....

$$(\vec{\vec{A}}\psi_0 - I)\frac{\partial\vec{U}_n}{\partial\xi} = \hat{\vec{H}}_n\{\vec{U}_0,\dots,\vec{U}_{n-1}\} - \vec{A}\nabla\vec{U}_{n-1} - \vec{\vec{A}}\psi_1\frac{\partial\vec{U}_{n-1}}{\partial\xi}. \tag{2.33}$$

The solution of the zeroth approximation is derived similarly to (2.8), provided Det $|A\vec{\psi}_0 - I| = 0$ is satisfied. Next, satisfying the orthogonality conditions for (2.31), the evolution equation governing $v_0(\xi,\vec{\chi})$ is easily obtained in similar fashion to (2.15) as

$$\frac{\partial v_0}{\partial s} = \frac{(\vec{\ell}\cdot\hat{\vec{F}})}{\lambda(\vec{\ell}\cdot\vec{r})} - \frac{(\vec{n}\cdot\vec{\psi}_1)}{(\vec{\ell}\cdot\vec{r})} \frac{\partial v_o}{\partial \xi} . \tag{2.34}$$

The next approximations, it is easily concluded, are similar to (2.16) and (2.17) and may be represented as

$$\frac{\partial v_1}{\partial s} = \hat{f}_1 - \frac{(\vec{n}\cdot\vec{\psi}_1)}{(\vec{\ell}\cdot\vec{r})} \frac{\partial v_1}{\partial \xi} \tag{2.35}$$

$$\frac{\partial v_n}{\partial s} = \hat{f}_n - \frac{(\vec{n}\cdot\vec{\psi}_1)}{(\vec{\ell}\cdot\vec{r})} \frac{\partial v_n}{\partial \xi} \tag{2.36}$$

However, the equations governing v_i are ambiguous due to the unknown function $\vec{\psi}_1$. Let us introduce the following transformation of independent variables

$$\eta = \xi - \frac{(\vec{n}\cdot\vec{\psi}_1)}{(\vec{\ell}\cdot\vec{r})} ds; \quad s' = s. \tag{2.37}$$

Introducing (2.37) into the equations (2.34) - (2.36) we arrive at the sequence of the equations

$$\partial v_0/\partial s' = \hat{f}_0, \tag{2.38}$$

$$\partial v_1/\partial s' = \hat{f}_1, \tag{2.39}$$

......

$$\partial v_2/\partial s' = \vec{f}_n. \tag{2.40}$$

In spite of the fact that these equations in terms of s' and η do not contain $\vec{\psi}_1$ explicitly, they cannot be solved without the ray trajectories being known. It is obvious that while the functions $\vec{f}_i$ depend upon the space vector $\vec{\chi}$, they must be recalculated in terms of s after the ray trajectories are found. Nevertheless, several physically important cases exist for which the solutions of (2.38) may be found without knowing the exact expressions for the ray trajectories [112].

Firstly, there is the class of separable operators

$$\hat{f}\{v_0, s\} = L(s)\hat{N}\{v_0\}, \tag{2.41}$$

where the explicit form of L(s) is unknown. Equation (2.38) with the right-hand side (2.41) governs, for example, the Riemann waves in an inhomogeneous medium or solitons in media with certain forms of inhomogeneity. The solution of equation (2.38) obtained either analytically or numerically may be written in the form

$$v_0 = v_0(\eta, \int L(s)ds). \tag{2.42}$$

Waves of almost constant profile are of the greatest interest. Therefore we associate the eikonal either with the maximum of v_0 (max $v_0 = \hat{v}_0$) or with another particular point of the profile. The equation of motion in terms of $\hat{v}_0$

$$\hat{\eta} = \eta[s, \hat{v}_0(s)], \tag{2.43}$$

determined by the solution (2.42) together with the transformation (2.37) and the condition ξ = const. gives the required expression for calculating the first term in the eikonal expansion

$$\hat{\eta} = \text{const} - \frac{(\vec{n}\cdot\vec{\psi}_1)}{(\vec{\ell}\cdot\vec{r})}\, ds, \tag{2.44a}$$

which may be represented as

$$\vec{n}\cdot\vec{\psi}_1 = -(\vec{\ell}\cdot\vec{r})\,\frac{d\vec{\eta}}{ds} \tag{2.44b}$$

thereby determining the function $\vec{\psi}_1$.

Secondly, another class of the solutions to the equation (2.38) may be found provided the medium parameters undergo slow changes in comparison with the characteristic lengths of the nonlinearity and dispersion etc. These conditions are usually more severe than the geometro-optical conditions with regard to the change of medium parameters in comparison with the wave length. The existence of the small parameter (the length of the nonlinearity over the scale parameter of the inhomogeneity) in the equation (2.38) makes possible the use of the averaging methods [39, 46, 81, 82, 116]. Generally speaking, equation (2.38) may be rewritten in an operator form

$$\hat{T}\{\eta, s, v_0, \mu\} = 0. \tag{2.45}$$

In the zeroth approximation with respect to μ the solution $v_0 = v_0^0(\eta, s, b_i)$ to this equation is easily found. Here b_i is a certain set of constants needed for describing the wave profile (amplitude, the averaged flow etc.). The parameters of the inhomogeneity as well as the cross-section of the ray-tube at this stage are considered constant. In this solution once more the profile maximum $\hat{v}_0^0$ is fixed, and then the corresponding coordinate $\hat{\eta}$ is determined and then it is easy to calculate the function $\vec{\psi}_1$ according to the expression (2.44b). However, in contrast to the solution (2.42), the function $v_0^0(\eta, s, b_i)$ is not a solution to the equation (2.45). In order to find this solution, we must determine the dependence of b_i upon the ray coordinate s. This dependence may be determined from the condition for nonsecularity of the correction to v_0^0. The correction v_0^1 is governed by the equations in variational derivatives

$$\frac{\delta \vec{T}}{\delta v_0^0} v_0^1 = \frac{\delta \vec{T}}{\delta \mu} . \tag{2.46}$$

The condition for nonsecularity of the correction v_0^1, as above, is the orthogonality condition of $\delta\vec{T}/\delta\mu$ to the eigenfunctions Y of the conjugate operator to $\delta\hat{T}/\delta v_0^0$

$$\langle Y \xrightarrow[\delta\mu]{\delta\hat{T}} 0. \tag{2.47}$$

Here <> denotes averaging over a wave period or over an infinite interval depending on the form of the unperturbed solution. Since $\delta\vec{T}/\delta\mu$ contains derivatives from the constants b_i, the conditions (2.47) are actually the equations determining the dependence of the amplitude (or its derivative) upon the parameters of the inhomogeneity. To find solutions to these equations is not an easy task; nevertheless in special cases (in the case of nondissipative media, for example) the amplitude equation degenerates into the principle of the conservation of energy or of energy flux. Thus the second term $\vec{\psi}_1$ in the eikonal gradient expansion is determined by equation (2.44) and the constants b_i involved in it are determined by the solution of (2.47).

<u>Example 2.2</u>. Let us consider a situation involving the separable operators (2.41) - the problem of nonlinear acoustics analysed in Section 2.1 (Example 2.1). In contrast to that analysis, we shall use here a scheme for propagation along nonlinear rays. Omitting the details we present here the final evolution equation

$$\frac{\partial p}{\partial s} + \left(\frac{c}{2}\,\phi_1 + \frac{\alpha}{\bar{\rho}c^3}\,p\right)\frac{\partial p}{\partial \xi} + \frac{c\bar{\rho}}{2S}\,\frac{d}{ds}\left(\frac{S}{c\bar{\rho}}\right)p = 0. \qquad (2.48)$$

This corresponds to equation (2.26) in Example 2.1 and has been derived by making use of the equation (2.35). Here $\vec{\psi}_1$ is replaced by the function $\phi_1 = 2\vec{\psi}_0\vec{\psi}_1$, where $\vec{\psi}_0^2 = c^{-2}$. The equation (2.48) may be simplified provided we replace the pressure p with the variable W

$$W = \{S/c\bar{\rho}\}^{1/2}\;p. \qquad (2.49)$$

This means that W^2 is proportional to the density of the energy flux. Now equation (2.48) may be rewritten as

$$\frac{\partial W}{\partial s} + (1/2\;\phi_1 + qW)\,\frac{\partial W}{\partial \xi} = 0. \qquad (2.50)$$

Here q is a parameter taking into account the changes in medium properties and ray-tube characteristics

$$q = \alpha\{\bar{\rho}c^5 S\}^{-1/2}. \tag{2.51}$$

Let us introduce the variables (2.37) into equation (2.50), noticing that $(\vec{\ell}\cdot\vec{r}) = 1$, $\vec{\psi}_0 = \vec{n}/c$. After solving the equation obtained in such a way, and fixing the maximum of the profile $W = p\{S/c\bar{\rho}\}^{1/2}$, we obtain for the expression determining ϕ_1

$$\phi_1 = -\frac{2q}{c}\hat{W} = -\frac{2\alpha}{\bar{\rho}c^4}\hat{p}. \tag{2.52}$$

Thus the full system of equations governing the nonlinear field and rays has the form

$$\nabla(\nabla\varphi/S|\nabla\varphi|) = 0 \tag{2.53a}$$

$$\nabla\varphi = \frac{\vec{n}}{c}\left(1 - \frac{\alpha\hat{p}}{\bar{\rho}c^2}\right), \tag{2.53b}$$

$$\hat{p}\{S/c\bar{\rho}\} = \text{const}, \tag{2.53c}$$

$$\frac{\partial p}{\partial s} + \frac{\alpha}{\bar{\rho}c^3}(p - \hat{p})\frac{\partial p}{\partial \xi} + \frac{c\bar{\rho}}{2S}\frac{d}{ds}\left(\frac{S}{c\bar{\rho}}\right)p = 0. \tag{2.53d}$$

System (2.53) is closed. The rays are determined for an amplitude $\hat{p}$ which changes due to the nonlinearity and the inhomogeneity. Here the ray equations do not depend upon the solution of the evolution equation, but the rays are essentially nonlinear. We shall have that in the case of linear rays (Section 2.1) the ray equations do depend upon the solution. Once the rays have been constructed, the wave field may then be calculated by means of evolution equations. The analogous equations may also be derived for shock waves. In this case, however, the ray equations and the evolution equations are coupled [31, 115].

Example 2.3. Let us consider now another situation corresponding to another class of solutions to the equation (2.38) [61, 85, 99]. Nonlinear waves in a dispersive medium are now under consideration. In this case equation (2.38) takes the form

$$\frac{\partial W}{\partial s} + \left(\frac{c}{2}\,\phi_1 + qW\right)\frac{\partial W}{\partial \xi} + \beta\,\frac{\partial^3 W}{\partial \xi^3} = 0, \tag{2.54}$$

i.e. it is the celebrated Korteweg-de Vries equation. Here we have used the notations analogous to the notations of equation (2.50). The coefficients c, q and β are, as a matter of fact, functions of the ray coordinate s. We shall introduce the transformation (2.37) into (2.54) and seek the solution to this equation in the form of a series expansion. The condition (2.46) for nonsecularity of the first correction gives us conservation of energy for a soliton

$$SE = \text{const}, \tag{2.55}$$

where E is the full energy of the soliton per unit area of its front. Since the solution is a stationary wave it is easy to fix the maximum point of the soliton which permits the determination of the ray correction ϕ_1

$$\phi_1 = -\frac{2}{3}\frac{q}{c}\hat{W}. \tag{2.56}$$

Taking into account that the energy of a soliton depends upon its amplitude ($E \sim W^{3/2}$) we now represent the full system of equations governing the nonlinear rays for the propagating soliton

$$\nabla(\nabla\varphi/S|\nabla\varphi|) = 0 \tag{2.57a}$$

$$\nabla\varphi = \frac{\vec{n}}{c}\left(1 - q\,\frac{c}{3}\,\hat{W}\right), \tag{2.57b}$$

$$\hat{W}^{3/2}S = \text{const}. \tag{2.57c}$$

Here the ray equations and the evolution equations are once again separated. Several solutions to the system (2.57) have been analysed elsewhere [61, 85].

The nonlinear rays used above have been determined with the accuracy of only one additional nonlinear term. We shall present here a modified asymptotic procedure in order to make corrections to the eikonal for every power ε^n. Such a procedure permits the construction of an asymptotic solution close to the exact solution at the intervals greater than ε^{-1}. The method of multiple scales is used and, similarly to (1.35), the new variables

$$\xi = \varphi(\vec{r}) - t, \quad \vec{X}_1 = \varepsilon r, \ldots, \vec{X}_n = \varepsilon^n \vec{r}, \tag{2.58}$$

are introduced. Substituting (2.58) into (2.1) we arrive at the system

$$(\vec{A} \sum_{k=0}^{n} \varepsilon^k \vec{\psi}_k - I) \frac{\partial \vec{U}}{\partial \xi} = \varepsilon \hat{\vec{F}} - \sum_{k=1}^{n} \varepsilon^k \vec{A} \frac{\partial \vec{U}}{\partial \vec{X}_k} . \tag{2.59}$$

Next we shall use the expansion of the eikonal gradient, and the solution to system (2.59) will be sought in the form of the asymptotic expansion (2.2). For the n-th approximation equation (2.59) yields

$$(\vec{A}\vec{\psi}_0 - I) \frac{\partial \vec{U}_n}{\partial \xi} = \hat{\vec{H}}_n \{\vec{U}_0, \ldots, \vec{U}_{n-1}\} - \\ - \vec{A}\vec{\psi}_n \frac{\partial \vec{U}_0}{\partial \xi} - \sum_{k=1}^{n} \vec{A} \frac{\partial \vec{U}_{n-k}}{\partial \vec{X}_k} . \tag{2.60}$$

Similarly to (1.40) we assume the eigensolutions U_k starting from k = 1 equal to zero. Hence, introducing the coordinate s_k $(\vec{A}\partial/\partial \vec{X}_k = \vec{A}\cdot\vec{n}\partial/\partial s_k)$ and satisfying the orthogonality conditions, we obtain for the variable $v = v_0 = \vec{\ell}\vec{U}_0$ the following equations in the powers $\varepsilon^1, \varepsilon^2, \ldots, \varepsilon^k$

$$\frac{\partial v}{\partial s_1} = \hat{f}_0(v) - (\vec{n}\vec{\psi}_1) \frac{\partial v}{\partial \xi} , \tag{2.61}$$

$$\frac{\partial v}{\partial s_2} = \hat{f}_1(v) - (\vec{n}\vec{\psi}_2) \frac{\partial v}{\partial \xi} - \frac{\partial v_{10}}{\partial s_1} , \tag{2.62}$$

......

$$\frac{\partial v}{\partial s_k} = \hat{f}_{k-1}(v) - (\vec{n}\vec{\psi}_k) \frac{\partial v}{\partial \xi} - \sum_{j=1}^{k-1} \frac{\partial v_{k-j,0}}{\partial s_j} . \tag{2.63}$$

Here v_{j0} denotes the forced solution where the normalization $\vec{\ell}\cdot\vec{r} = 1$ has been taken into account. The system given above differs from the system

(2.34)-(2.36) in that every equation here contains the corresponding correction of $\vec{\psi}_k$. Similarly to (1.41) we have

$$\frac{\partial v}{\partial s} = \frac{\partial v}{\partial s_1} + \varepsilon \frac{\partial v}{\partial s_2} + \dots . \tag{2.64}$$

Thus the equations (2.61)-(2.63) turn into one equation

$$\frac{\partial v}{\partial s} + (\widehat{\vec{n}\vec{\psi}}) \frac{\partial v}{\partial \xi} = f - \frac{\partial v_b}{\partial s}, \tag{2.65a}$$

$$f = \hat{f}_0 + \sum_{j=1}^{k} \varepsilon^j \hat{f}_j, \tag{2.65b}$$

$$v_b = \sum_{j=1}^{k-1} \varepsilon^j v_{j0}, \tag{2.65c}$$

$$\hat{\vec{\psi}} = \sum_{j=1}^{k} \varepsilon^j \vec{\psi}_j. \tag{2.65d}$$

Introducing the transformation (c.f. the transformation (2.37))

$$\eta = \xi - \int (\widehat{\vec{n}\vec{\psi}})ds, \quad s' = s \tag{2.66}$$

we arrive at the equation (c.f. the equation (2.38))

$$\frac{\partial v}{\partial s'} = f - \frac{\partial v_b}{\partial s'} . \tag{2.67}$$

This equation once again does not explicitly contain the ray characteristics and therefore, similar to (2.44), the following expression may be used

$$(\vec{n}\cdot\hat{\vec{\psi}}) = - \frac{d\hat{\eta}}{ds} . \tag{2.68}$$

2.3 The estimation of nonlinear effects in a near-caustic zone

The applicability of nonlinear geometrical methods is based on the assumption that both the medium and the wave-field parameters undergo smooth changes.

If these parameters change rapidly diffraction effects occur. Such a situation may happen at the singular points of the ray structure, including caustics. The wave-fields containing the discontinuities at the near-caustic zone have been investigated thoroughly and the differential equations governing the field in these zones are known [4, 59, 117, 118]. If the wave-field does not contain any discontinuities then the linear approach may be used. Here we assume that at the near-caustic zone the diffraction effects are stronger than the nonlinear ones. Hence at the first approximation we shall limit ourselves only to diffraction effects.

We shall justify this approach by using an example of a nonlinear wave equation. As the starting point let us choose equation (2.18) which after some obvious transformations may be written in the form

$$\Delta p - \frac{1}{c^2}\frac{\partial^2 p}{\partial t^2} - \frac{\nabla\rho}{\rho}\ p = \frac{\alpha}{\rho c^4}\frac{\partial^2 p^2}{\partial t^2} . \tag{2.69}$$

The solution to this equation is sought in terms of the series expansion

$$p = p_1 \exp i\omega t + \varepsilon p_2 \exp 2i\omega t + \dots . \tag{2.70}$$

Using the well-known perturbation scheme we arrive at the sequence

$$\Delta p_1 + k^2 p_1 - \nabla\rho/\rho\ \ \nabla p_1 = 0, \tag{2.71a}$$

$$\Delta p_2 + 4k^2 p_2 - \nabla\rho/\rho\ \nabla p_2 = - \frac{4\alpha}{\rho c^2} k^2 p_1^2. \tag{2.71b}$$

Here $k = \omega/c(\vec{r})$. We shall now introduce the new variables $\psi_{1,2} = p_{1,2}/\rho^{1/2}$. Neglecting the terms proportional to $\Delta\rho$ and $(\nabla\rho)^2$, the equations (2.71) yield

$$\nabla\psi_1 + k^2\psi_1 = 0, \tag{2.72a}$$

$$\nabla\psi_2 + 4k^2\psi_2 = -4\alpha\omega^2\psi_1^2/\rho^{1/2}c^4. \tag{2.72b}$$

Further we shall consider the following problem [12]: the plane $z = 0$

divides the space into the homogeneous ($z < 0$) and the inhomogeneous ($z > 0$) halfspaces; the sound velocity dependence on the space coordinate at $z > 0$ is given by

$$c^2 = c_0^2 (1 - az)^{-1}, \tag{2.73}$$

where c_0^2, a are constants. The plane wave emanates from the "depth" of the homogeneous halfspace and will be partly reflected from, partly transmitted through the interface. The expression (2.73) permits us to find the exact solution to equations (2.72). The total field at $z < 0$ is the sum of the incident and reflected waves and may be presented as

$$\psi_1(x,z) = Z_1(z) \exp i\bar{\xi}x, \tag{2.74a}$$

$$\psi_2(x,z) = Z_2(x) \exp 2i\bar{\xi}x, \tag{2.74b}$$

$$\bar{\xi} = k_0 \sin \Theta_0, \tag{2.74c}$$

$$Z(z) = \exp(ik_0 z \cos \Theta_0) + V(\Theta_0)\exp(-ik_0 z \cos \Theta_0). \tag{2.74d}$$

Here Θ_0 is the incident angle, the amplitude of the incident wave is unity and $V(\Theta)$ is the reflection coefficient. For the transmitted wave ($z > 0$) when equations (2.72) contain $k(z) = k_0 c_0/c(z)$ we shall transform the variables according to the expressions

$$\tau = \tau_0 + z/H, \tag{2.75a}$$

$$\tau_0 = -H^2 k_0^2 \cos^2\Theta_0, \quad H = (ak_0^2)^{-1/3}. \tag{2.75b}$$

Introducing (2.75) into equation (2.72) we obtain

$$\frac{d^2 Z_1}{d\tau^2} - \tau Z_1 = 0, \tag{2.76a}$$

$$\frac{d^2 Z_2}{d\tau^2} - 4\tau Z_2 = \text{const} - AZ_1^2. \tag{2.76b}$$

Obviously the Airy functions $Z(\tau)$ are the solutions to the equation (2.76a). The most important phenomenon here is the fact that the waves change their direction and may have a certain turning point. This turning point is labelled as $\tau = 0$. We shall now investigate the behaviour of the second harmonic at $\tau = 0$. The solution to the equation (2.76b) has the form [51]

$$Z_2(\tau) = -v(4^{1/3}\tau) \int_{-\tau}^{\infty} u(4^{1/3}\Theta)v^2(\Theta)d\Theta +$$

$$+ u(4^{1/3}\tau) \int_{-\tau}^{\infty} v(4^{1/3}\Theta)v^2(\Theta)d\Theta. \qquad (2.77)$$

This solution is expressed through the linearly-independent solutions of u and v to the equation (2.76a). The Wronskian of these functions is equal to unity. The right-hand side of equation (2.76b) is assumed to contain $z_1 = v(\tau)$.

Two integrals are needed in order to calculate $Z_2(\tau)$. Let us express them in the following form ($W_1 = u$, $W_2 = v$)

$$J_{1,2} = \int_{-\tau}^{-\tau_k} W_{1,2}(4^{1/3}\Theta)v^2(\Theta)d\Theta + \int_{-\tau_k}^{0} W_{1,2}(4^{1/3}\Theta)v^2(\Theta)d\Theta, \qquad (2.78)$$

where τ_k corresponds to the caustic. The first integral describes the field far from the caustic. Let us estimate the value of the second integral that determines the amplitude of the second harmonic in a rather close near-caustic zone starting at $\tau = \tau_k$ up to, say, $\tau = 0$. We note that the Airy function decays rapidly at $\tau > 0$. Here we have

$$J_2^1 = \int_{-\tau_k}^{0} u(4^{1/3}\Theta)v^2(\Theta)d\Theta, \qquad (2.79a)$$

$$J_2^2 = \int_{-\tau_k}^{0} v(4^{1/3}\Theta)v^2(\Theta)d\Theta. \qquad (2.79b)$$

Expanding the Airy functions into a series and using the transformation $t = -\tau$ we obtain

$$J_2^1 = \int_0^{\tau_k} \cos\left(\frac{4}{3}\, t^{3/2} + \frac{\pi}{4}\right)\sin^2\left(\frac{2}{3}\, t^{3/2} + \frac{\pi}{4}\right) \frac{dt}{4^{1/3} t^{3/4}}, \tag{2.80a}$$

$$J_2^2 = \int_0^{\tau_k} \sin\left(\frac{4}{3}\, t^{3/2} + \frac{\pi}{4}\right)\sin^2\left(\frac{2}{3}\, t^{3/2} + \frac{\pi}{4}\right) \frac{dt}{4^{1/3} t^{3/4}}. \tag{2.80b}$$

In order to estimate the value of τ_k we assume the numerator in the integrals is constant (several units). Then

$$J_2^1 \sim J_2^2 \sim 4\tau_k^{1/4}. \tag{2.81}$$

This coincides with the maximum field value $Z_2(\tau)$ at the caustic. Thus we have $Z_2(\tau)/Z_1(\tau) \sim 4\tau_k^{1/4}$. By making use of the notations (2.75), and taking into account the dimension of the near-caustic zone $z_k' = z_k - \tau_0 H$ that is of the wave-length order, we arrive at the following condition

$$M\left(\frac{z_k}{\lambda}\right)^{1/4} \left(\frac{a}{k_0}\right)^{1/2} << 1 \tag{2.82}$$

where the Mach number $M = p_1(0)/\rho c_0^2$. This gives a condition on the smallness of the second harmonic at the caustic.

It is obvious that the condition on short waves is rather weak, and the second harmonic at a caustic is small provided $M << 1$. The first integral in expression (2.77) may also be easily estimated and gives the possibility of demonstrating the growth of the second harmonic. Let us once more expand the Airy function into a series and change the sign of the argument. The result is

$$J_1^1 = \int_{\tau_k}^{\tau} \left\{\cos\left(\frac{4}{3}\, t^{3/2} + \frac{\pi}{4}\right) + \sin\left(\frac{8}{3}\, t^{3/2} + \frac{\pi}{4}\right) - \right.$$

$$\left. - \cos\frac{\pi}{4}\right\} \frac{dt}{2^{3/2} t^{3/4}}, \tag{2.83a}$$

$$J_1^2 = \int_{\tau_k}^{\tau} \left\{\sin(\frac{4}{3} t^{3/2} + \frac{\pi}{4}) - \cos(\frac{8}{3} t^{3/2} + \frac{\pi}{4}) + \right.$$

$$\left. + \sin \frac{\pi}{4}\right\} \frac{dt}{2^{3/2} t^{3/4}} . \tag{2.83b}$$

The main part of the integrals is determined by the constant terms in the numerators. Performing the integration (2.83) we arrived at the approximate solution

$$Z_2(\tau) = 2 \left\{\sin (\frac{4}{3} \tau^{3/2} + \frac{\pi}{4}) + \cos (\frac{4}{3} \tau^{3/2} + \frac{\pi}{4})\right\}. \tag{2.84}$$

Consequently, far from the caustic, the amplitude of the second harmonic remains constant. Since the amplitude of the first harmonic decreases, the relation $Z_2/Z_1 \sim Z^{1/4}$ (the correction to the linear solution) increases. Thus, at the near-caustic zone the nonlinear effects are small. Therefore, in the first approximation, the near-caustic zone may be calculated by means of the linear theory of diffraction, but up to and after the caustic the nonlinear effects must be calculated by the formulae given above (see also [84, 86, 105]).

Let us consider now a continuous wave at a caustic. The incident periodic wave of arbitrary form v_{inc} may be expanded into a Fourier series around the ray coordinate at several wave-lengths from the caustic

$$v_{inc}(\xi) = \sum_{n=1}^{\infty} (a_n \cos n\xi + b_n \sin n\xi) \tag{2.85a}$$

$$\xi = t - \int ds/c(s). \tag{2.85b}$$

It is known from the exact solution that every spectral component is shifted in phase up to $-\pi/2$ after reflection from a caustic due to diffraction. Therefore the reflected wave-field has the form

$$v_{ref}(\xi) = \sum_{n=1}^{\infty} (-a_n \sin n\xi + b_n \cos n\xi). \tag{2.86}$$

Performing an inverse Fourier-transform in order to find the coefficients

a_n, b_n we obtain a relation between the incident and reflected waves in the form of the integral Hilbert transform

$$v_{ref}(\xi) = \frac{1}{4}\int_{-\pi}^{\pi} v_{inc}(\xi')\,\mathrm{ctg}\,\frac{\xi' - \xi}{2}\,d\xi'. \tag{2.87}$$

Here we have used the generalized functions $\sum_{n=1}^{\infty} \sin n\xi = \frac{1}{2}\,\mathrm{ctg}\,\frac{\xi}{2}$ for calculating the sine-series.

The form of the reflected wave depends explicitly on the form of the incident wave. If the incident wave to the caustic is harmonic then the reflected wave is also harmonic. If due to the nonlinear effects the incident wave is a N-wave with the amplitude v^+, then the reflected wave according to the expression (2.87) has a logarithmic singularity

$$v_{ref}(\xi) = v^+ \ln \{2|\sin \xi/2|^{1/2\pi}\}. \tag{2.88}$$

Here the linear approach is no longer correct. However, from the physical viewpoint this singularity may be eliminated and estimates for the bounded amplitude may be found. This problem is analysed in detail elsewhere [10, 83, 98].

Similarly the field of strong pulses (waves) at the near-front region may also be calculated [28, 88, 98]. Here, however, a certain physical phenomenon not mentioned above may occur. Suppose we have a pulse signal where the wave-lengths of the low-frequency components are bigger than the characteristic dimension of the ray-tube. These components do not "recognize" the caustic and so a certain precursor may be formed. This phenomenon has been analysed rather well elsewhere [43, 53, 111]. For the high-frequency components of the signal the Hilbert transform is used, which in this case is given by

$$v_{ref}(\xi) = \frac{1}{\pi}\int_{-\pi}^{+\pi} \frac{v_{inc}(\xi')}{\xi - \xi'}\,d\xi'. \tag{2.89}$$

Thus at the first approximation for waves that do not contain any discontinuity at the near-caustic region the linear theory is applicable. The corrections at higher approximations may be found by the perturbation method.

2.4 The asymptotic scheme for wave-beams

Let us consider in detail the asymptotic scheme for constructing the evolution equations in case of wave-beams. The notion of wave-beams is further used in the sense of certain fields with parameters (the amplitude, the wave-length etc.) which change slowly in the direction of propagation and rapidly in the transverse direction. In linear theory wave-beams are often described by using the so-called "paraxial" approximation [7]. The scheme presented here is actually a nonlinear variant to the paraxial approximation.

Let us once again return to the system of equations (2.1) and assume for the sake of simplicity that all the elements of the matrix A do not depend on the space coordinates. Bearing in mind a rather general class of wave-beams at $\varepsilon \neq 0$ propagating along x-axis we shall introduce the new variable

$$u = Yv, \tag{2.90}$$

where Y is the square matrix formed by the linearly independent eigenvectors to the matrix A_x. The system (2.1) in terms of (2.90) yields

$$\frac{\partial v_i}{\partial t} + \lambda_i \frac{\partial v_i}{\partial x} + \sum_j \vec{B}_{ij} \nabla_\perp v_j = \varepsilon(Y^{-1}\hat{\vec{F}})_i = \varepsilon f_i, \tag{2.91a}$$

$$B = Y^{-1} \vec{A}_\perp Y, \; \nabla_\perp = \{\partial/\partial y, \partial/\partial z\}, \; A_\perp = \{A_y, A_z\}. \tag{2.91b}$$

Here λ_i are the eigenvalues to the matrix A_x and Y^{-1} is the inverse matrix to Y. The initial and boundary conditions for equation (2.91) are easily deduced using (2.90).

We shall consider here only the one-wave approximation (the multi-wave approximation is analysed similarly). As was mentioned above, wave-field changes along different directions are of different scales. Let the change along the propagation direction be f. We assume $f = \varepsilon\varphi$ and v depends on y and z in terms of ε: $y_H = \varepsilon y$, $z_H = \varepsilon z$. Such a presentation means that the wave-field varies faster in the transverse direction and is distorted slower along the propagation direction. It means that

$$\frac{\partial v_i}{\partial t} + \lambda_i \frac{\partial v_i}{\partial x} = - \varepsilon \sum_j \vec{B}_{ij} \nabla_{\perp H} v_j + \varepsilon^2 \varphi_i. \tag{2.92}$$

Now we shall apply the iterative procedure described in Section 1.1. The equation (2.92) yields in the zeroth approximation

$$\frac{\partial v_1}{\partial t} + \lambda_i \frac{\partial v_1}{\partial x} = 0, \tag{2.93a}$$

$$v_i = 0, \; i = 2,\ldots,n. \tag{2.93b}$$

In the first approximation we have

$$\frac{\partial v_1}{\partial t} + \lambda_1 \frac{\partial v_1}{\partial x} = - \varepsilon \vec{B}_{11} \nabla_{\perp H} v_1, \tag{2.94a}$$

$$\frac{\partial v_i}{\partial t} + \lambda_i \frac{\partial v_i}{\partial x} = \varepsilon \vec{B}_{i1} \nabla_{\perp H} v_1, \tag{2.94b}$$

and in the second approximation -

$$\frac{\partial v_1}{\partial t} + \lambda_1 \frac{\partial v_1}{\partial x} = -\vec{B}_{11} \nabla_{\perp H} v_1 - $$

$$- \varepsilon \sum_{i=2}^{n} \vec{B}_{1i} \nabla_{\perp H} v_i + \varepsilon^2 f_1\{v_1\}, \tag{2.95}$$

where v_i are determined beforehand from the equation (2.94b). Introducing the variables

$$\xi = x - \lambda_1 t, \tag{2.96a}$$

$$\xi_i = x - \lambda_i t, \quad i \neq 1, \tag{2.96b}$$

equation (2.94b) is easily integrated as

$$v_i = \varepsilon \int \vec{B}_{i1} \nabla_{\perp H} v_1(\xi)(\lambda_i - \lambda_1)^{-1} d\xi. \tag{2.97}$$

Thus the "multidimensional" term in equation (2.95) may be rewritten as

$$Q = - \varepsilon \sum_{i=2}^{n} \vec{B}_{1i} \nabla_{\perp H} v_i =$$

$$= \varepsilon^2 \sum_{i=2}^{n} (\vec{B}_{1i} \nabla_{\perp H})(\vec{B}_{i1} \nabla_{\perp H})(\lambda_i - \lambda_1)^{-1} \int v_1 d\xi. \qquad (2.98)$$

In the case of the two-dimensional problem

$$Q = \varepsilon^2 \sum_{i=2}^{n} \vec{B}_{1i} \vec{B}_{i1}(\lambda_i - \lambda_1)^{-1} \int \frac{\partial^2 v_1}{\partial y_H^2} d\xi =$$

$$= \varepsilon^2 q \int \frac{\partial^2 v_1}{\partial y_H^2} d\xi. \qquad (2.99)$$

Introducing the variables

$$\xi = x - \lambda t, \quad \tau = \varepsilon t \qquad (2.100)$$

equation (2.95) together with the expression (2.99) reads

$$\frac{\partial v_1}{\partial \tau} = - B_{11} \frac{\partial v_1}{\partial y_H} + \varepsilon q \int \frac{\partial^2 v_1}{\partial y_H^2} d\xi + \varepsilon f_1. \qquad (2.101)$$

We could simplify this equation by once more introducing the new variables

$$y' = y_H - B_{11}\tau, \quad \tau' = \varepsilon\tau . \qquad (2.102)$$

Now equation (2.101) may be rewritten (neglecting the primes)

$$\frac{\partial v_1}{\partial \tau} = q \int \frac{\partial^2 v_1}{\partial y^2} d\xi + f_1. \qquad (2.103)$$

Differentiating equation (2.103) with respect to ξ we arrive at the equation for the second approximation

$$\frac{\partial}{\partial \xi} \left\{ \frac{\partial v_1}{\partial \tau} - f_1 \right\} = q \frac{\partial^2 v_1}{\partial y^2} . \qquad (2.104)$$

If $f_1 = av_1 \partial v_1/\partial\xi$, then the equation (2.104) is called the Zabolotskaya-Khokhlov equation after the researchers who deduced it in order to describe wave-beams within the framework of nonlinear acoustics [120]. If

$$f_1 = av_1 \frac{\partial v_1}{\partial \xi} + \beta \frac{\partial^3 v}{\partial \xi^3}, \; q < 0, \tag{2.105}$$

then equation (2.104) is the Kadomtsev-Petviashvili equation [48]. In the linear case ($f_1 = 0$) and the monochromatic wave ($v_1 = v \exp ik\xi$) equation (2.104) becomes the parabolic equation

$$ik \frac{\partial v}{\partial \tau} = q \frac{\partial^2 v}{\partial y^2}, \tag{2.106}$$

which corresponds to the "paraxial approximation" [7].

The details of constructing evolution equations for higher approximations are given elsewhere [25] and will not be analysed here.

Finally we consider a problem arising in practical applications - how the constants q and B_{11} depend upon the medium properties, say in the case of an isotropic medium. We shall use here a property of linearized evolution equations. As indicated already in the Introduction, the linear part of an evolution equation may be deduced from the dispersion relation. In the case of an isotropic nondispersive medium the dispersion relation has the form

$$\omega = \lambda k = \lambda(k_x^2 + k_y^2)^{1/2}. \tag{2.107}$$

In wave-beams the condition $k_y \ll k_x$ is satisfied which permits representation of expression (2.107) in another form

$$\omega \sim \lambda k_x + \frac{1}{2}\lambda \, k_y^2 \, k_x^{-1}, \tag{2.108}$$

or, separating the quadratic part,

$$k_x(\omega - \lambda k_x) = \frac{1}{2}\lambda \, k_y^2. \tag{2.109}$$

Using the obvious correspondence (see the Introduction) $k_x \to -i\partial/\partial x$, $k_y \to -i\partial/\partial y$, $\omega \to i\partial/\partial t$, the expression (2.109) may be replaced by the evolution equation

$$\frac{\partial}{\partial x}\left(\frac{\partial}{\partial t} + \lambda \frac{\partial}{\partial x}\right)v = -\frac{\lambda}{2}\frac{\partial^2 v}{\partial y^2}. \quad (2.110)$$

Comparing the result with equation (2.104) at $f_1 = 0$ and taking all the transformations used above into account we arrive at

$$B_{11} = 0, \quad q = -\frac{1}{2}\lambda. \quad (2.111)$$

It is easy to prove that $B_{ii} = 0$ for arbitrary i provided the medium is isotropic. This calculation of constants for the equation (2.104) was proposed by Kadomtsev and Petviashvili [48].

2.5 Wave-beams in weakly inhomogeneous media

Suppose the properties of a medium depend weakly on the space coordinates. Then the matrix A contains space-dependent functions and the trajectory of the paraxial wave-beam is curved. There is no general theory treating this case, therefore we shall consider a special case of nonlinear acoustics further and demonstrate the paraxial approximation using an example of the nonlinear wave equation. In terms of pressure it reads

$$\frac{1}{c^2}\frac{\partial^2 p}{\partial t^2} - \Delta p + \varepsilon \frac{\nabla\rho}{\rho}\nabla p = \varepsilon\phi, \quad (2.112)$$

where the term responsible for the inhomogeneity is separated. Here both the density ρ and the sound velocity c are assumed to depend on the space coordinates. The right-hand side of this equation, besides nonlinear terms, may also contain terms describing dissipation, relaxation etc. [23, 90].

Once again, the transformation from the multi-wave situation into the one-wave approximation needs the ray variable $\xi = \varphi(\vec{r}) - t$, and the wave-field is assumed to change slowly along the ray s and rapidly along the transverse direction m

$$p = p(\xi, \varepsilon s, \varepsilon^{1/2} m) \quad (2.113)$$

Substituting (2.113) into equation (2.112) we obtain

$$[c^2(\nabla\varphi)^2 - 1]\frac{\partial^2 p}{\partial\xi^2} +$$

$$+ c^2 \frac{\partial}{\partial\xi}[2\nabla\varphi\nabla p + (\nabla\varphi - \frac{\nabla\rho}{\rho}\nabla\varphi)p + \Phi] + c^2\frac{\partial^2 p}{\partial m^2} = 0. \tag{2.114}$$

The eikonal gradient in (2.114) determines the direction of the wave-beam but the cross-section of the ray-tube is not used here (c.f. Section 2.1). Instead of it, only one supporting ray exists here. This corresponds formally to a unit cross-section in expressions (2.25) or (2.28)

$$\Delta\varphi = \frac{d}{ds}(\frac{1}{c}) \tag{2.115}$$

and the details of this are to be found elsewhere [4]. Taking this expression into account and writing

$$\Pi(m,s) = c^2(\nabla\varphi)^2 - 1, \tag{2.116}$$

we arrive at the equation

$$\frac{\partial}{\partial\xi}[\frac{\partial p}{\partial s} + \frac{c\rho}{2}\frac{d}{ds}(\frac{1}{c\rho})p + \Phi] = \frac{c}{2}\frac{\partial^2 p}{\partial m^2} + \frac{\Pi}{2c}\frac{\partial^2 p}{\partial\xi^2} = 0. \tag{2.117}$$

The biggest difficulties are associated with the transformation of the last term in equation (2.117). Here we present the dependence of Π on the ray and transverse coordinates for the simplest case. The propagation time along the supporting ray is obviously $\varphi = \int_0^s ds/c(s,o)$. According to a well-known formula [4] we have in a curvilinear system of coordinates

$$(\nabla\varphi)^2 = q_{11}^{-1}(s,m)c^{-2}(s,o), \tag{2.118}$$

where q_{11} is the component of the matrix tensor equal to the square of the Lamé parameter. The simplest result is obtained for a plane beam. Here we get

$$q_{11} = (1 + \frac{m}{R(s)}), \tag{2.119}$$

where R(s) is the radius of curvature of the ray. Substituting these expressions into (2.116), expanding the velocity c(s,m) into a series in the transverse coordinate, and taking into account the dependent of the curvature on the sound velocity [1]

$$R = c/\left\{\frac{dc}{dm}\right\}_{m=0}, \qquad (2.120)$$

we obtain finally to the order of ε

$$\Pi(s,m) = \frac{m^2}{c}\left\{\frac{\partial^2 c}{\partial m^2}\right\}_{m=0}. \qquad (2.121)$$

Thus the evolution equation describing the paraxial beam in a weakly inhomogeneous media has the final form

$$\frac{\partial}{\partial\xi}\hat{L}(p) + \frac{c}{2}\frac{\partial^2 p}{\partial m^2} + \frac{m^2}{2c^2}\left\{\frac{\partial^2 c}{\partial m^2}\right\}_{m=0}\frac{\partial^2 p}{\partial\xi^2} = 0, \qquad (2.122a)$$

$$\hat{L}(p) = \frac{\partial p}{\partial s} + \frac{c\rho}{2}\frac{d}{ds}\left(\frac{1}{c\rho}\right)p + \Phi. \qquad (2.122b)$$

We point out that in the case of propagation along the gradient of the inhomogeneity we have $\Pi = 0$, and therefore equation (2.122) turns into equation (2.104) with variable coefficients. If the function ϕ is proportional to $p\partial p/\partial\xi$, then equation (2.122) turns into a modification of the Zabolotskaya-Khokhlov equation. Certain solutions to this equation obtained by the perturbation method, are known [93].

3 The wave–guides and evolution equations

The wave propagation processes in which the transverse structure of profiles is inhomogeneous occurs in several completely different physical situations. The wave-guide-like motion may be connected with the existence of free boundaries (for example, the surface and the bottom of the ocean, protective shells etc.) or with the stratified structure of media (the vertical structure of hydrological fields in the ocean). A typical example of the inhomogeneous structure is the propagation of gravitational waves on the surface of water. Emphasis must be placed on the fact that the celebrated Korteweg-de Vries equations already mentioned in Chapters 1 and 2 was originally derived for waves in shallow water as far back as 1895. The techniques for constructing evolution equations in wave-guides is not worked out in detail, particularly when compared with the results for quasi-one-dimensional systems. Therefore, besides the mathematical scheme, full attention will also be paid to constructing the evolution equation in specific situations.

3.1 The Galerkin procedure for eliminating the "non-wave" coordinate

The approaches given above in Chapters 1 and 2 are not directly applicable for wave-guides, whose governing equations also contain functions of the "non-wave" coordinate and its derivatives. The structure of the propagating wave-mode depends directly upon this coordinate. Therefore it is appropriate to work out a mathematical procedure which enables the elimination of the non-wave coordinate, and as a result the transformation of the wave-guide equations into equations containing only the "wave" coordinates (i.e. the coordinates along which waves propagate). Such a procedure was mentioned in the Introduction while describing long waves on the surface of water. According to this procedure the velocity potential was expanded into a series along the vertical coordinate and the boundary value problem was transformed to the two-dimensional equations for the shallow water. A much more general procedure may be constructed on the basis of the Galerkin approximation that uses a series expansions with respect to the eigenmodes.

This procedure will be described here in a more detailed way.

The essence of the Galerkin procedure is traditionally explained by using the example of an inhomogeneous linear equation written in the operator form [63]

$$\hat{L}u - F = 0 \tag{3.1}$$

subject to certain homogeneous boundary conditions. Let us choose a certain sequence of basic functions (modes) M_i. For the sake of simplicity we assume that these functions depend only upon the non-wave coordinate, say upon the vertical coordinate z. In addition we assume that these functions are differentiable as many times as needed and satisfy the boundary conditions. Then the function

$$u_n(x,y,z,t) = \sum_{i=1}^{n} v_i(x,y,t)M_i(z), \tag{3.2}$$

where v_i are the arbitrary functions, also satisfies all of the boundary conditions. According to the Galerkin method, the functions v_i are determined from the condition that the left-hand side of equation (3.1), after the substitution of u_n into it instead of u, becomes orthogonal to the functions $M_1,\ldots,M_n$ [63]. This approach gives us a linear system in terms of functions v_i

$$\sum_{i=1}^{n} v_i\langle M_j, \hat{L}M_i\rangle = \langle M_j, F\rangle, \tag{3.3}$$

where $\langle f,\varphi\rangle$ stands for the scalar product. If the operator L is nonlinear but the boundary conditions, as previously, are homogeneous, then the principle of using the orthogonality condition for determining the functions v_i remains the same. Thus we obtain a rather simple procedure for eliminating the "non-wave" coordinate. The estimation of accuracy in this case is however less clear. In the particular case of ordinary integro-differential equations, the problem has been investigated [63], while in the case of nonlinear partial differential equations only certain assumptions may be made. It is natural to suppose that in the case of weak nonlinearity the wave-field expansion with respect to the eigenmodes determined by the corresponding linear problem, will be successful. Here the number of eigenmodes is limited due to the physical conditions of their generation, and to the interaction between the different modes. It is obvious that, generally speaking, the eigenmodes depend on the frequency and the expansion (3.2) has meaning only for the spectrum. If the dispersion is weak, then the dependence of mode structure upon the frequency may be neglected. This assumption leads us directly to an expansion in the form of (3.2), and the equations governing $v_i(x,y,t)$ are easily found. As a result we find a finite number of equations in terms of v_i, and the methods described in Chapters 1 and 2 may be successfully used.

Example 3.1 Let us consider the application of the Galerkin procedure [79, 80] in the dynamics of internal waves in a stratified incompressible liquid. We start from the equations

$$\frac{\partial w}{\partial z} + \operatorname{div} \vec{u} = 0, \tag{3.4}$$

$$\frac{\partial \rho'}{\partial t} + w \frac{d\rho_0}{dt} = - w[\frac{\partial \rho'}{\partial z} + \vec{u}\nabla\rho'] = S_1, \tag{3.5}$$

$$\frac{\partial p'}{\partial z} + g\rho' = - \rho[w \frac{\partial w}{\partial z} + \vec{u}\nabla w] - \rho \frac{\partial w}{\partial t} = S_2, \tag{3.6}$$

$$\rho_0 \frac{\partial \vec{u}}{\partial t} + \nabla p' = -[\rho' \frac{\partial \vec{u}}{\partial t} + \rho w \frac{\partial \vec{u}}{\partial z} + \rho(\vec{u}\nabla)\vec{u}] = S_3. \tag{3.7}$$

Here w and $\vec{u}$ are the vertical and the horizontal particle velocity components, respectively; p is the pressure, ρ - the density while $\rho_0(z)$ is the unperturbed density, ρ' and p' are the perturbed density and pressure, respectively. The operator ∇ acts only in the horizontal (x,y) plane.

Beside the nonlinear terms, the right-hand side of equation (3.6) also contains the "nonhydrostatic" term $\rho\partial w/\partial t$ which for long waves (in comparison with the ocean depth) is always small. This fact is easily proved by analyzing the properties of the equations (3.4)-(3.7). Further we shall make the usual assumptions valid for internal waves: (i) the assumption of a solid cover (i.e. w = 0 at the free surface); (ii) the Boussinesq assumption (the density $\rho_0(z)$ is assumed to be always constant when it is not differentiated). Thus we shall assume that all terms on the right-hand side of the system

(3.4)-(3.7) are small.

The solution to the system (3.4)-(3.7) is sought in the form of an expansion containing the eigenmodes M_i of the corresponding linear non-dispersive system. The latter follows from the system above with $S_1 = S_2 = S_3 = 0$. The structure of the solution is as follows

$$w = \sum_m M_m(z)\tilde{w}_m(x,y,t), \tag{3.8}$$

$$\vec{u} = \sum_m C_{1m} \frac{dM_m}{dz} \tilde{\vec{u}}(x,y,t), \tag{3.9}$$

$$\rho' = \sum_m C_{2m} M_m(z) N^2(z) \tilde{\rho}_m(x,y,t), \tag{3.10}$$

$$p' = \sum_m C_{3m}\rho_0 \frac{dM_m}{dz} \tilde{p}_m(x,y,t). \tag{3.11}$$

Here $N(z) = (-gd(\ln \rho_0)/dz)^{1/2}$ is the Brunt-Väisälä frequency (note that the z-axis is directed upwards and $d\rho_0/dz < 0$) and C_{1m}, C_{2m}, C_{3m} are the separation constants. Their choice is governed, for example, by the dimensionality of new variables with a tilde in the expressions (3.9)-(3.11). The function $M_m(z)$ is governed by the boundary problem

$$\frac{dM_m(z)}{dz^2} + \frac{N^2(z)}{C^2} M_m = 0 \tag{3.12}$$

subject to the homogeneous conditions

$$M_n(0) = M_m(H) = 0 \tag{3.13}$$

at the upper and lower boundaries of the liquid layer under consideration. Here C_m is the eigenvalue determined from the boundary problem and it has actually the meaning of the mode propagation velocity at the long-wave limit; H is the ocean depth. It should also be remarked that the functions M_m satisfy the orthogonality conditions

$$\langle N^2 M_k, M_\ell \rangle = \langle \frac{dM_k}{dz}, \frac{dM_\ell}{dz} \rangle = 0, \quad k \neq \ell \tag{3.14}$$

Let us now apply the Galerkin procedure for determining the functions $\tilde{w}, \tilde{\vec{u}}, \tilde{\rho}, \tilde{p}$. First of all we note that the system of equations (3.4)-(3.7) is actually a set of equations of type (3.1), therefore the Galerkin procedure may be applied independently to every equation. We shall multiply the equations (3.4) and (3.7) by dM_m/dz, and equations (3.5) and (3.6) by M_m. Integrating now with respect to z from z = 0 to z = H, we arrive at the following system

$$\tilde{w}_m + C_{1m} \operatorname{div} \tilde{\vec{u}}_m = 0, \tag{3.15}$$

$$C_{2m} \frac{\partial \tilde{\rho}_m}{\partial t} - \frac{\rho}{g} \tilde{w}_m = \frac{\langle M_m, S_1 \rangle}{\langle N^2 M^2 \rangle}, \tag{3.16}$$

$$g C_{2m} \tilde{\rho}_m - \rho_0 \frac{C_{3m}}{C_m^2} \tilde{p}_m = \frac{\langle M_m, S_2 \rangle}{\langle N^2 M_m^2 \rangle}, \tag{3.17}$$

$$C_{1m} \frac{\partial \tilde{\vec{u}}}{\partial t} + C_{3m} \nabla \tilde{p}_m = \frac{\langle dM_m/dz, \vec{S}_3 \rangle}{\rho_0 \langle (dM_m/dz)^2 \rangle}. \tag{3.18}$$

This system no longer depends on the z coordinate and is of order 4N, where N is the number of terms in the series expansions with respect to the eigenmodes. In the linear case this system is separable into N independent systems of equations of the fourth order. The choice of N within the framework of the Galerkin procedure is not determined (see above). The physical background dictates the interactions only between several modes provided the nonlinearity is weak. These modes are related by the resonance expressions (the conditions of synchronism) while the wave-fields of other modes subject to certain initial conditions remain small and can be calculated afterwards by the perturbation method. As a result, the system obtained is of a comparatively low order and contains only the horizontal coordinates and derivatives with respect to them. Such systems were analysed in Chapters 1 and 2. Although the Galerkin procedure is not rigorously established, the resulting equations are physically clearly admissible since the usual approximation ideas such as the "one-wave" character, for example, are not used. Therefore these equations can govern many processes in wave propagation and diffraction problems with inhomogeneous transverse structures. Here we should like to point out that when the "one-wave" assumption is taken into account, the resulting simplified system (3.15)-(3.18) evolution equations coincides exactly with the equations of the first asymptotic expansion. This problem will be discussed in detail later on (see Section 3.2).

The system of equations (3.15)-(3.18) governing the dynamics of internal gravitational waves in stratified liquids may be transformed into another form that is typical of surface waves in homogeneous liquids [79, 80]. For this purpose it is convenient to introduce instead of w the variable ζ that measures the decline of the liquid line where the density is constant (isopycns), from the unperturbed level. Obviously at this surface we have

$$w = \frac{\partial \zeta}{\partial t} + (\vec{u}\nabla)\zeta. \qquad (3.19)$$

Linearizing this expression, we find an expansion for ζ with respect to the modes of the unperturbed problem

$$\zeta = \sum_m M_m(z)\eta_m(x,y,t). \qquad (3.20)$$

Applying the Galerkin procedure and taking into account equations (3.15)-(3.18) we arrive at the system (for details see [43, 45, 46])

$$\frac{\partial \eta}{\partial t} + \operatorname{div}\,[(H + \frac{1}{2}S\eta)\tilde{\vec{u}}] = 0, \qquad (3.21)$$

$$\frac{\partial \tilde{\vec{u}}}{\partial t} + \frac{c^2}{H}\nabla\eta + S[(\tilde{\vec{u}}\nabla)\tilde{\vec{u}} - \frac{1}{2H}\frac{\partial}{\partial t}(\eta\tilde{\vec{u}})] + DH\frac{\partial^2\eta}{\partial t^2} = 0, \qquad (3.22)$$

provided the interactions between modes are absent. The separation constants in (3.9)-(3.11) have been chosen in such a way that the dimensions of η and $\tilde{\vec{u}}$ coincide with those of ζ and $\vec{u}$. In equations (3.21) and (3.22) the following notation has been used

$$S = \frac{H\langle(dM/dz)^3\rangle}{\langle(dM/dz)^2\rangle}, \quad D = \frac{\langle M^2\rangle}{H^2\langle(dM/dz)^2\rangle} \qquad (3.23)$$

where the index m is neglected. It must be emphasized that all the nonlinear terms in equations (3.21) and (3.22) have one and the same coefficient S determined by the structure of the corresponding mode. In the case N(z) =

const. this parameter turns out to be zero, i.e. the influence of the nonlinearity (the nonlinear "selfaction") arises only in the next approximation. The quadratic nonlinearity arises only then if the Boussinesq approximation [75] or the assumption of the solid cover are neglected. Introducing now the variables

$$\zeta = H + S\eta, \quad \vec{v} = S\tilde{\vec{u}} - \frac{S^2}{2H}\eta\vec{u} \tag{3.24}$$

into the equations (3.21) and (3.22) we get to the same order of approximation

$$\frac{\partial \xi}{\partial t} + \operatorname{div}(\xi\vec{v}) = 0, \tag{3.25}$$

$$\frac{\partial \vec{v}}{\partial t} + (\vec{v}\nabla)\vec{v} + \bar{g}\nabla\xi + \frac{c^2 DH}{2}\nabla\Delta\xi = 0. \tag{3.26}$$

These equations coincide with the Boussinesq equations for the surface waves in shallow water already presented in the Introduction. Here $\bar{g} = c^2/H$ is the reduced acceleration due to gravity. However, it must be clearly understood that in reality the parameter S may have various numerical values and even a sign dependent on the character of the stratification. In this sense the situations for internal waves may be more varied when compared with surface waves.

Thus long internal and surface waves are described by the same system of equations, and consequently their dynamics must be similar. Making use of the methods described in Chapter 1, for the travelling plane two-dimensional wave, an evolution equation is easily deduced having the form of the Korteweg-de Vries equation. In terms of $v = \tilde{u}_x$ it reads

$$\frac{\partial v}{\partial t} + c\frac{\partial v}{\partial x} + \frac{3}{2}Sv\frac{\partial v}{\partial x} + \frac{cDH^2}{2}\frac{\partial^3 v}{\partial x^3} = 0. \tag{3.27}$$

For quasiplanes waves we obtain in similar way the Kadomtsev-Petviashvili equation

$$\frac{\partial}{\partial x}\left\{\frac{\partial v}{\partial t} + c\frac{\partial v}{\partial x} + \frac{3}{2}Sv\frac{\partial v}{\partial x} + \frac{cDH^2}{2}\frac{\partial^3 v}{\partial x^3}\right\} = \frac{c}{2}\frac{\partial^2 v}{\partial y^2}. \tag{3.28}$$

The solutions to these equations are known [110].

The Example above was about the Galerkin procedure applied directly to the equations of motion. When the Lagrangian or the Hamiltonian for the equations is known, the procedure needs a wave-field expansion with the respect to eigenmodes like

$$u = \sum_m \int dk_x dk_y u^m_{k_x,k_y} M_m(z) \exp [i(\omega_m t - k_x x - k_y y)], \tag{3.29}$$

which must be introduced into the Lagrangian (or the Hamiltonian). The next step - the variation of the Lagrangian (or the Hamiltonian) with respect to the spectral amplitudes $u^m_{k_x,k_y}$ according to the rules given in Section 1.4 - gives the final result. The special features of wave-guide problems appears only in the matrix coefficients of interaction that depend upon the integrals of the modes $M_m(z)$. The integration itself is carried out automatically in the course of calculating the Lagrangian (the Hamiltonian). Further the equations for the spectral amplitudes are easily transformed into evolution equations (c.f. Section 1.4).

Thus the Galerkin procedure for the elimination of the "non-wave" coordinate permits the construction of simplified equations containing only the horizontal coordinates (the "wave" coordinates) and derivatives with respect to them. Bearing in mind the minimization of the order of the resulting equations, it is recommended in case of the weak nonlinearity and dispersion to choose the modes of the linear nondispersive problem as the basic functions. The number of modes depends on the number of generated waves and on the resonance conditions which determine the interactions between the modes. Further the simplified equations may be easily transformed into the evolution equations (see Chapters 1 and 2).

3.2 The asymptotic method for the one-wave approximation

The Galerkin procedure together with the rules of choosing the eigenmodes from the corresponding linear non-dispersive problem as the basic functions may be clearly justified by the asymptotic methods in case of a one-wave process. Here we shall present an asymptotic expansion scheme for a comparatively general system of the following form

$$\frac{\partial}{\partial t} \hat{L}_1 \{z\}\vec{U} + \frac{\partial}{\partial x} \hat{L}_2 \{z\}\vec{U} = \varepsilon\vec{F}, \qquad (3.30)$$

where $\hat{L}_1$ and $\hat{L}_2$ are linear matrix operators on z which for the sake of exposition are considered to be differential. The wave variables must be determined subject to the linear homogeneous boundary conditions

$$\hat{L}_3\{z\}\vec{U} = 0, \quad z = z_i, \quad i = 1,2. \qquad (3.31)$$

The case of nonlinear boundary conditions will be discussed later. At $\varepsilon = 0$ the solution to the system (3.30) subject to the conditions (3.31) is obviously found in the form of nondispersive linear waves

$$\vec{U} = \vec{M}(z)v(x - ct), \qquad (3.32)$$

where v is the scalar function in the moving frame x - ct and M(z) is the vector of eigenfunctions corresponding to one of the modes of the following boundary problem

$$(\hat{L}_2 - c\hat{L}_1)\vec{M} = 0, \quad \vec{L}_3\vec{M} = 0, \quad z = z_i \qquad (3.33)$$

with the eigenvalue c. Clearly, this solution generalizes the expression (1.28) making it acceptable for the wave-guide problems.

At $\varepsilon = 0$ we shall introduce the variables

$$\xi = x - ct, \quad \tau = \varepsilon t \qquad (3.34)$$

and consider the initial value problem (c.f. Section 1.2 for the boundary value problem). The solution is sought in the form of the asymptotic series

$$\vec{U}(t,x,z) = \vec{U}_0(z,\xi,\tau) + \vec{U}_1(z,\xi,\tau) + \dots . \qquad (3.35)$$

In the new variables (3.34), system (3.30) reads

$$(\hat{L}_2 - c\hat{L}_1) \frac{\partial\vec{U}}{\partial\xi} = \varepsilon\vec{F} - \varepsilon\vec{L}_1 \frac{\partial\vec{U}}{\partial\tau} . \qquad (3.36)$$

Introducing (3.35) into this system and equating the coefficients of like powers in ε, we obtain the equations for determining $\vec{U}_n$. In particular, in the zeroth approximation we find $\vec{U}_0$

$$\vec{U}_0 = \vec{M}(z)v(\xi,\tau), \tag{3.37}$$

where $v(\xi,\tau)$ in the accuracy of this approximation is arbitrary. In the first approximation we obtain the inhomogeneous system for $\vec{U}_1$

$$(\hat{L}_2 - c\hat{L}_1)\frac{\partial \vec{U}_1}{\partial \xi} = \vec{F}\{z,\xi,\vec{U}_0\} - \hat{L}_1\vec{M}\frac{\partial v}{\partial \tau}. \tag{3.38}$$

Here the form of the boundary conditions is not changed

$$\hat{L}_3\vec{U}_1 = 0, \quad z = z_i. \tag{3.39}$$

This inhomogeneous boundary problem (3.38) and (3.39) may be solved provided the orthogonality conditions

$$\langle\vec{M}^*\{\vec{F} - \frac{\partial v}{\partial \tau}\hat{L}_1\vec{M}\}\rangle = 0 \tag{3.40}$$

are satisfied. Here $\vec{M}^*$ is the eigenvector conjugate to (3.33). This condition determines the evolution equation in terms of v

$$\frac{\partial v}{\partial \tau} = \Phi_1\{\xi,v\} = \frac{\langle\vec{M}^*, \vec{F}\rangle}{\langle\vec{M}^*, \hat{L}_1\vec{M}\rangle}. \tag{3.41}$$

When condition (3.40) is satisfied, the solution to equation (3.38) may be written in the form

$$\vec{U}_1 = \vec{U}_{1f} + \vec{M}(z)v_1(\xi,\tau), \tag{3.42}$$

where $\vec{U}_{1f}$ is the forced solution to (3.38) determined by $\vec{U}_0$ and v_1 is an arbitrary (up to this approximation) scalar function.

The next approximation gives an inhomogeneous system of equations in terms of $\vec{U}_2$ with the right-hand side containing the term $\hat{L}_1\vec{M}\partial v_1/\partial\tau$ and known functions of v. Once more, the compatibility conditions for this inhomogeneous boundary value problem give as an evolution equation for the second approximation

$$\frac{\partial v_1}{\partial \tau} = \Phi_2(\xi, v, v_1). \tag{3.43}$$

The equations of the next approximations are constructed similarly. The asymptotic scheme just described may be modified in order to obtain an evolution equation in terms of only one function v, the form of which depends on the approximation number. Such a modification may be constructed by introducing the multi-scale coordinates $\xi, \tau_1 = \varepsilon t$, $\tau_2 = \varepsilon^2 t$, $\tau_3 = \varepsilon^3 t, \ldots$ similarly, as was done in Section 1.2. Another way to obtain such an evolution equation is to seek it directly in the form

$$\frac{\partial v}{\partial \tau} = \Phi_1 \{\xi, v\} + \varepsilon\Phi_2\{\xi, v\} + \ldots, \tag{3.44}$$

where the functionals Φ_i must be determined from the compatibility conditions for the inhomogeneous boundary value problem at every power of ε. Here, obviously, all functions v_i in the expressions of type (3.42) ought to be equal to zero. It is also not difficult to generalize this approach for smoothly inhomogeneous (along the propagation plane) and slowly non-stationary media, but we shall not discuss this here.

Here is the proper place to prove that the equations of the first approximation obtained by using the asymptotic method coincide with those obtained here through the Galerkin procedure, provided the basic functions chosen are the eigenmodes for the corresponding linear nondispersive problem. For this purpose the wave-field is represented in the form of the mode expansion

$$U_i = \sum_n \bar{M}_{in} u_{in}(x, t). \tag{3.45}$$

Contrary to (3.37), this expansion is not unique for every index i because u_{in} does not correspond to one travelling wave, but only to one mode. Introducing expansion (3.45) into the initial system (3.30) we obtain

$$\sum_{j,n} (\hat{L}_{1ij}\bar{M}_{jn} \frac{\partial u_{jn}}{\partial t} + \hat{L}_{2ij}\bar{M}_{jn} \frac{\partial u_{jn}}{\partial x}) = \varepsilon F_i. \tag{3.46}$$

According to the Galerkin procedure we multiply this system by $\bar{M}_{im}$ and integrate over z within a layer. The resulting system of equations no longer contains the coordinate z

$$\sum_{i,n} (\langle\bar{M}_{im}, \hat{L}_{1ij}\bar{M}_{jn}\rangle \frac{\partial u_{jn}}{\partial t} + \langle\bar{M}_{im}, \hat{L}_{2ij}\bar{M}_{jn}\rangle \frac{\partial u_{jn}}{\partial x}) =$$

$$= \varepsilon\langle\bar{M}_{im}, F_i\rangle. \tag{3.47}$$

In case of a one-mode process, the system (3.47) is simplified and neglecting the index n = m we arrive at the following system

$$\sum_{j} (\langle\bar{M}_i, \hat{L}_{1ij}\bar{M}_j\rangle \frac{\partial u_j}{\partial t} + \langle\bar{M}_i, \hat{L}_{2ij}\bar{M}_j\rangle \frac{\partial u_j}{\partial x}) = \varepsilon\langle\bar{M}_i, F_i\rangle. \tag{3.48}$$

Finally the one-dimensional system of equations in terms of u_j may be written as

$$A \frac{\partial \vec{U}}{\partial t} + B \frac{\partial \vec{U}}{\partial x} = \varepsilon\vec{F}, \tag{3.49}$$

which was actually analyzed in Chapter 1. The only difference is that the matrix A is not a unit matrix. This is not an essential difference, and therefore we now present the resulting evolution equation without detailed discussion. It has the traditional (c.f. Chapter 1) form

$$\frac{\partial v}{\partial \tau} = \Phi_1 = \frac{\vec{\ell}\,\vec{F}}{\vec{\ell}\,A\,\vec{r}}, \tag{3.50}$$

where $\vec{\ell}$ and $\vec{r}$ are the left and right eigenvectors for the matrix B - cA (c is a certain eigenvalue obtained for $\varepsilon = 0$ and has the dimensions of a velocity). Remembering the expression for determining the matrices A and B, the right-hand side of equation (3.50) may be expressed as

$$\Phi_1 = \frac{\sum_i \int \ell_i M_i F_i dz}{\sum_{i,j} \int \ell_i \bar{M}_i \hat{L}_{1ij} \bar{M}_j r_j dz} . \quad (3.51)$$

Substituting $u_i = vr_i$ into expression (3.45) we obtain the relationship between the components of the eigenvector corresponding to one wave M_i in the epxression (3.37) and the functions $\bar{M}_i$ obtained from (3.45) by separating the variables

$$M_i = \bar{M}_i r_i . \quad (3.52)$$

Similarly it may be shown that $\ell_i \bar{M}_i$ coincides with the components of the eigenvectors to the conjugate boundary system M_i^*.

This means that the functional Φ_1 can be rewritten as

$$\Phi_1 = \frac{\langle \bar{M}^*, \vec{F} \rangle}{\langle \bar{M}^*, \hat{L}_1 \bar{M} \rangle} , \quad (3.53)$$

which completely coincides with the right-hand side of (3.41). Thus the coincidence of the evolution equations obtained by the Galerkin procedure and by the asymptotic method in the first approximation is proved. This fact is another proof of the advantages making a suitable choice of the basic functions used in the Galerkin procedure.

Example 3.2. Let us apply the asymptotic method to the two-dimensional wave process in a stratified liquid [64]. This process was analyzed in Section 3.1 using the Galerkin procedure. Introducing the flux function ψ, so that

$$u = -\frac{\partial \psi}{\partial z} , \quad w = \frac{\partial \psi}{\partial x} , \quad (3.54)$$

equations (3.4)-(3.7) are reduced to the standard form (3.30). In terms of ψ they become:

$$\frac{\partial \rho'}{\partial t} + \frac{d\rho_0}{dz}\frac{\partial \psi}{\partial x} = \frac{\partial \psi}{\partial z}\frac{\partial \rho'}{\partial x} - \frac{\partial \psi}{\partial x}\frac{\partial \rho'}{\partial z} , \quad (3.55a)$$

$$\frac{\partial}{\partial z}\left(\rho\frac{d}{dt}\frac{\partial\psi}{\partial z}\right) + \frac{\partial}{\partial x}\left(\rho\frac{d}{dt}\frac{\partial\psi}{\partial x}\right) + g\frac{\partial\rho'}{\partial x} = 0. \qquad (3.55b)$$

This system must be solved subject to the boundary conditions

$$\frac{\partial\psi}{\partial x} = 0 \quad \text{at} \quad z = 0, H. \qquad (3.56)$$

In order to determine the possible small parameters it is convenient to introduce the dimensionless variables

$$z' = z/H, \; x' = x/\lambda, \quad t' = ct/\lambda, \qquad (3.57a)$$

$$\psi' = \psi/u_0H, \; \rho_0' = \rho_0/\rho^*, \quad (\rho')' = gH\rho'/cu_0\rho^*. \qquad (3.57b)$$

Here ρ^* is the average density of the liquid, u_0 is the characteristic amplitude of the horizontal particle velocity, λ is the wave length and c - its propagation velocity. Neglecting the primes, we arrive at the system

$$\frac{\partial\rho}{\partial t} + \frac{gH}{c^2}\frac{d\rho_0}{z}\frac{\partial\psi}{\partial x} = \varepsilon\left(\frac{\partial\psi}{\partial z}\frac{\partial\rho}{\partial x} - \frac{\partial\psi}{\partial x}\frac{\partial\rho}{\partial z}\right), \qquad (3.58)$$

$$\frac{\partial^2}{\partial z\partial t}\left(\rho_0\frac{\partial\psi}{\partial z}\right) + \frac{\partial\rho}{\partial x} =$$

$$= \varepsilon\frac{\partial}{\partial z}\left\{\rho_0\frac{\partial\psi}{\partial z}\frac{\partial^2\psi}{\partial z\partial x} - \frac{\partial\psi}{\partial x}\frac{\partial^2\psi}{\partial z^2} - \frac{c^2}{gH}\rho\frac{\partial^2\psi}{\partial z\partial t}\right\} -$$

$$- \mu\rho_0\frac{\partial^3\psi}{\partial t\partial x^2} + \varepsilon^2\frac{c^2}{gH}\frac{\partial}{\partial z}\left\{\rho\left(\frac{\partial\psi}{\partial z}\frac{\partial^2\psi}{\partial z\partial x} - \frac{\partial\psi}{\partial x}\frac{\partial^2\psi}{\partial z^2}\right)\right\} -$$

$$- \varepsilon\mu\left\{\frac{c^2}{gH}\frac{\partial}{\partial x}\left(\rho\frac{\partial^2\psi}{\partial t\partial x}\right) + \rho_0\frac{\partial}{\partial x}\left(\frac{\partial\psi}{\partial x}\frac{\partial^2\psi}{\partial z\partial x} - \frac{\partial\psi}{\partial z}\frac{\partial^2\psi}{\partial x^2}\right)\right\} +$$

$$+ \varepsilon^2\mu\frac{\partial}{\partial x}\left\{\rho\left(\frac{\partial\psi}{\partial z}\frac{\partial^2\psi}{\partial x^2} - \frac{\partial\psi}{\partial x}\frac{\partial^2\psi}{\partial z\partial x}\right)\right\}. \qquad (3.59)$$

The right-hand sides of these equations contain the nonlinear and dispersion parameters

$$\varepsilon = u_0/c, \quad \mu = H^2/\lambda^2, \tag{3.60}$$

respectively. Further we shall discuss the wave processes characterized by the small parameters of one and the same order, i.e. $\varepsilon \sim \mu << 1$. It must be noted that the dispersion may also be weak when $\mu >> 1$. This case will be discussed in detail in Section 3.3. Generally speaking, an evolution equation may also be constructed in the case that the parameters ε, μ are not small. Such a case is analysed elsewhere [13, 14]. It is clear, however, that it is an exceptional case. Here, working with small parameters ε, μ, the equations (3.58), (3.59) are of the following schematical form

$$\frac{\partial \rho}{\partial t} + \frac{gH}{c^2}\frac{d\rho_0}{dz}\frac{\partial \psi}{\partial x} = \varepsilon Q_0, \tag{3.61}$$

$$\frac{\partial}{\partial z}\left(\rho_0 \frac{\partial^2 \psi}{\partial z \partial t}\right) + \frac{\partial \rho}{\partial x} = \varepsilon Q_1 + \varepsilon^2 Q_2 + \varepsilon^3 Q_3. \tag{3.62}$$

Using the notation

$$\vec{U} = \begin{vmatrix} \rho \\ \psi \end{vmatrix} \quad \hat{L}_1 = \begin{vmatrix} 1 & 0 \\ 0 & \frac{\partial}{\partial z}\left(\rho_0 \frac{\partial}{\partial z}\right) \end{vmatrix}, \tag{3.63a}$$

$$\hat{L}_2 = \begin{vmatrix} 0 & \frac{gH}{c^2}\frac{d\rho_0}{dz} \\ 1 & 0 \end{vmatrix}, \quad \vec{F} = \begin{vmatrix} Q_0 \\ Q_1 + \varepsilon Q_2 + \varepsilon^2 Q_3 \end{vmatrix}, \tag{3.63b}$$

the equations (3.61), (3.62) are easily rewritten in the standard form (3.30). Obviously, at $\varepsilon = 0$ the system of equations (3.61), (3.62) has a solution in the form of travelling waves (in the x-direction)

$$\psi(t,x,z) = M(z)v(x - ut), \tag{3.64a}$$

$$\rho = \frac{gH}{uc^2}\frac{d\rho_0}{dz}\psi, \tag{3.64b}$$

where v is an arbitrary function determined by the character of the generated internal waves, $M(z)$ and u are the eigenfunctions and the corresponding eigenvalues, respectively, determined from the boundary problem

$$\frac{d}{dz}\left(\rho_0 \frac{dM}{dz}\right) - \frac{gH}{u^2c^2}\frac{d\rho_0}{dz} M = 0, \tag{3.65a}$$

$$M(0) = M(1) = 0. \tag{3.65b}$$

If $\varepsilon \neq 0$, then the solution is sought in the form of the asymptotic expansion (3.35) with the deformed variables

$$\xi = x - ut, \quad \tau = \varepsilon t. \tag{3.66}$$

The first term in this expansion is given by formulae (3.64), and v satisfies the evolution equation

$$\frac{\partial v}{\partial \tau} = \Phi_1\{\xi, v\} + \varepsilon\Phi_2\{\xi, v\} + \dots . \tag{3.67}$$

Substituting (3.35), (3.64)-(3.67) into equations (3.61), (3.62) we obtain the inhomogeneous boundary system

$$-u\frac{\partial U_{1n}}{\partial \xi} + \frac{gH}{c^2}\frac{d\rho_0}{z}\frac{\partial U_{2n}}{\partial \xi} = R_n + \frac{gH}{uc^2} M\Phi_n, \tag{3.68a}$$

$$\frac{\partial U_{1n}}{\partial \xi} - u\frac{\partial}{\partial z}\left(\rho_0 \frac{\partial^2 U_{2n}}{\partial z \partial \xi}\right) = L_n + \frac{gH}{uc^2}\frac{d\rho_0}{\partial z} M\Phi_n. \tag{3.68b}$$

$$R_1 = C_0(\varepsilon = 0), \tag{3.69a}$$

$$R_m = \frac{1}{(m-1)!}\frac{\partial^{m-1}}{\partial \varepsilon^{m-1}} Q_0 \ (\varepsilon = 0), \quad m \geq 2, \tag{3.69b}$$

$$L_1 = Q_1 \ (\varepsilon = 0), \tag{3.69c}$$

$$L_2 = \frac{\partial}{\partial \varepsilon} Q_1 \ (\varepsilon = 0) + Q_2 \ (\varepsilon = 0), \tag{3.69d}$$

$$L_3 = \frac{1}{2}\ \frac{\partial^2}{\partial\varepsilon^2} Q_1\ (\varepsilon = 0) + \frac{\partial}{\partial\varepsilon} Q_2\ (\varepsilon = 0) + Q_2\ (\varepsilon = 0), \tag{3.69e}$$

$$L_m = \frac{1}{(m-1)!} \frac{\partial^{m-1}}{\partial\varepsilon^{m-1}} Q_1(\varepsilon = 0) + $$

$$+ \frac{1}{(m-2)!} \frac{\partial^{m-2}}{\partial\varepsilon^{m-2}} Q_2\ (\varepsilon = 0) + $$

$$+ \frac{1}{(m-3)!} \frac{\partial^{m-3}}{\partial\varepsilon^{m-3}} Q_3\ (\varepsilon = 0),\ m \geqq 4. \tag{3.69f}$$

The boundary conditions (3.56) yield for the corrections

$$U_{2n} = 0, \quad z = 0, 1. \tag{3.70}$$

The inhomogeneous boundary value problem (3.68), (3.69) subject to the boundary conditions (3.70) is solvable provided the orthogonality condition (the boundary problem (3.65) is self-conjugate)

$$\int_0^1 dzM(z) \left\{\frac{2gH}{u^2c^2} \frac{d\rho_0}{dz} M\Phi_n + \frac{R_n}{u} + L_n\right\} = 0 \tag{3.71}$$

is satisfied. This condition determines the summands Φ_n in the evolution equation (3.67)

$$\Phi_n = -\frac{u^2c^2}{gH} \frac{\int_0^1 dz\ M(z)\ [L_n + R_n/u]}{\int_0^1 dz[M(z)]^2\ d\rho_0/dz} . \tag{3.72}$$

In particular, in the first approximation (of power ε) equation (3.67) yields

$$\frac{\partial v}{\partial\tau} + \alpha v \frac{\partial v}{\partial\xi} + \beta \frac{\partial^3 v}{\partial\xi^3} = 0, \tag{3.73a}$$

$$\alpha = \frac{3}{2} \frac{\int_0^1 \rho_0 (dM/dz)^3 dz}{\int_0^1 \rho_0 (dM/\partial z)^2 dz}, \tag{3.73b}$$

$$\beta = \frac{u}{2} \frac{\int_0^1 \rho_0 \, M^2 dz}{\int_0^1 \rho_0 (dM/dz)^2 dz}. \tag{3.73c}$$

It is not surprising that the result is the Korteweg-de Vries equation deduced in Section 3.1 by using the Galerkin procedure. The Boussinesq assumption permitting the calculation of integrals with ρ_0 assumed to have a constant value is not used here. If this assumption were to be used, the coincidence in coefficients is obvious. A more detailed derivation of the Korteweg-de Vries equation describing internal waves in liquid media with a horizontal inhomogeneity of density is given in [92].

Finally in this Section we shall discuss nonlinear boundary conditions. This means that conditions (3.31) have a certain right-hand side

$$\hat{L}_j\{z\}\vec{U} = \varepsilon\vec{R}, \quad z = z_i, \quad i = 1,2. \tag{3.74}$$

The approximation of the zeroth order is naturally not influenced, but the boundary condition for the first correction in the asymptotic expansion becomes inhomogeneous (c.f. the formula (3.39))

$$\hat{L}_3\{z\}\vec{U}_1 = \vec{R}(\xi,\vec{U}_0\}. \tag{3.75}$$

There exist standard procedures which may be applied to the boundary problem (3.38) together with the boundary conditions (3.75) in order to get the corresponding homogeneous problem. Let us present the solution to (3.38) in the form

$$\vec{U}_1 = \vec{U}_{10} + \vec{U}_{1N}, \tag{3.76}$$

where U_{1N} is a function chosen to satisfy only the boundary conditions (3.75).

Then the function $\vec{U}_{10}$ will be determined by the following homogeneous boundary problem

$$(\hat{L}_2 - c\hat{L}_1) \frac{\partial \vec{U}_{10}}{\partial \xi} = \vec{F}\{z, \xi, \vec{U}_0\} -$$

$$- (\hat{L}_2 - c\hat{L}_1) \frac{\partial \vec{U}_{1N}}{\partial \xi} - \hat{L}_1 \vec{M} \frac{\partial v}{\partial \tau}, \qquad (3.77a)$$

$$\hat{L}_3 \vec{U}_{10} = 0. \qquad (3.77b)$$

The compatibility conditions for this problem give immediately the evolution equation of the first order

$$\frac{\partial v}{\partial \tau} = \frac{\langle \bar{M}^*, [\vec{F} - (\hat{L}_2 - c\hat{L}_1)\partial \vec{U}_{1N}/\partial \xi] \rangle}{\langle \bar{M}^*, \hat{L}_1 \bar{M} \rangle}. \qquad (3.78)$$

A detailed example of such an asymptotic approach to internal waves in the ocean with a free upper surface (without the solid cover approximation) and nonlinear boundary conditions is given in [69, 75]. The free surface means that the coefficients of the equation (3.73) are different, i.e. the nonintegral terms are added, which is characteristic of non-zero boundary conditions.

Thus the wave distortion in media with an inhomogeneous transverse structure may be described using asymptotic procedures which are the natural generalizations of methods described in Chapters 1 and 2. The difficulties here, as a rule, are due to the complications in solving the boundary value problems in order to find the eigenmodes in specific physical situations, and are not of principal significance.

3.3 Wave-guides of complicated structures

The results presented in the previous Section were obtained for the systems of type (3.30) which describe nondispersive waves at the zeroth approximation. Generally speaking, the boundary problems may be of rather specific type and are not always reducible to the systems of type (3.30). For example the boundary problem for surface waves in an inhomogeneous incompressible liquid governed by equations (0.7)-(0.10) is definitely not of type (3.30), while the derivatives with respect to time exist only in the boundary conditions. Bearing in mind such complicated situations, the asymptotic procedure must be modified. The general scheme of modification is not known, and therefore the analysis must be further restricted to a special example. This example concerns the dynamics of internal waves in an ocean of the infinite depth (c.f. the same problem in an ocean of the finite depth in Section 3.2). It should be pointed out that the dispersion parameter given by the expression (3.60) is not small for this problem, but to the contrary tends to the infinity. Let us assume that the middle layer with thickness 2h is stratified, while the upper and the lower layers are both homogeneous. Then the waves in the middle layer are, as before, governed by equations (3.55), and outside this layer by the equation

$$\frac{\partial}{\partial t}\left(\frac{\partial^2}{\partial x^2}+\frac{\partial^2}{\partial y^2}\right)\psi + \left(\frac{\partial\psi}{\partial z}\frac{\partial}{\partial x}-\frac{\partial\psi}{\partial x}\frac{\partial}{\partial z}\right)\left(\frac{\partial^2}{\partial x^2}+\frac{\partial^2}{\partial y^2}\right)\psi = 0. \tag{3.79}$$

The following boundary conditions must be satisfied at the interfaces

$$\{\psi\} = 0, \quad \left\{\frac{\partial\psi}{\partial z}\right\} = 0, \quad z = \pm h + \zeta_{1,2}. \tag{3.80}$$

Here $\zeta_{1,2}$ are the displacements of the upper and lower interfaces of the stratified layer (pycnocline) and z = 0 corresponds to the central plane of the layer. These conditions correspond to the situation when the density at the interfaces is continuous [57]. The boundary conditions at infinity

$$\psi \to 0, \quad z \to \pm\infty \tag{3.81}$$

must certainly be added. Once more, let us use dimensionless variables

$$x' = x/\lambda, \; z' = z/h, \; t' = ct/\lambda, \; \varepsilon = a/h, \tag{3.82a}$$

$$\zeta' = \zeta/a, \; \psi' = \psi/\varepsilon ch, \; \rho_0' = \rho_0/\rho^*, \; (\rho')' = \rho/\varepsilon\rho^*, \tag{3.82b}$$

where a is the amplitude of the displacement at the interfaces, c is as usual the propagation velocity, and λ is the wavelength. Substituting (3.82) into (3.55) and (3.79) we obtain for the middle layer

$$\frac{\partial \rho}{\partial t} + \frac{d\rho_0}{dz}\frac{\partial \psi}{\partial x} = \varepsilon Q_0, \tag{3.83a}$$

$$\frac{\partial}{\partial z}\left(\rho_0 \frac{\partial^2 \psi}{\partial z \partial t}\right) + \frac{\partial \rho}{\partial x} = \varepsilon Q_1 + \varepsilon^2 Q_2 + \delta^2 Q_3 + $$

$$+ \delta^2 \varepsilon Q_4 + \delta^2 \varepsilon^2 Q_5 , \tag{3.83b}$$

$$Q_0 = \frac{\partial \rho}{\partial z}\frac{\partial \psi}{\partial x} - \frac{\partial \rho}{\partial x}\frac{\partial \psi}{\partial z} , \tag{3.84a}$$

$$Q_1 = - \left\{\frac{\partial}{\partial z}\left(\rho_0 \frac{\partial \psi}{\partial z}\frac{\partial^2 \psi}{\partial z \partial x}\right) - \frac{\partial}{\partial z}\left(\rho_0 \frac{\partial \psi}{\partial x}\frac{\partial^2 \psi}{\partial z^2}\right) + \right.$$

$$\left. + \frac{\partial}{\partial z}\left(\rho_0 \frac{\partial^2 \psi}{\partial z \partial t}\right)\right\}, \tag{3.84b}$$

$$Q_2 = - \frac{\partial}{\partial z}\left\{\rho_0 \left(\frac{\partial \psi}{\partial z}\frac{\partial^2 \psi}{\partial z \partial x} - \frac{\partial \psi}{\partial x}\frac{\partial^2 \psi}{\partial z^2}\right)\right\}, \tag{3.84c}$$

$$Q_3 = - \frac{\partial}{\partial x}\left(\rho_0 \frac{\partial^2 \psi}{\partial x \partial t}\right), \tag{3.84d}$$

$$Q_4 = - \frac{\partial}{\partial x}\left\{\rho_0 \left(\frac{\partial \psi}{\partial z}\frac{\partial^2 \psi}{\partial x^2} - \frac{\partial \psi}{\partial x}\frac{\partial^2 \psi}{\partial x \partial z}\right) + \rho \frac{\partial^2 \psi}{\partial x \partial t}\right\}, \tag{3.84e}$$

$$Q_5 = - \frac{\partial}{\partial x}\left\{\rho_0 \left(\frac{\partial \psi}{\partial z}\frac{\partial^2 \psi}{\partial x^2} - \frac{\partial \psi}{\partial x}\frac{\partial^2 \psi}{\partial x \partial z}\right)\right\}, \tag{3.84f}$$

and outside of it

$$\frac{\partial}{\partial t}\left(\frac{\partial^2}{\partial x^2} + \frac{\partial^2}{\partial y^2}\right)\psi_{1,2} + $$

$$+ \varepsilon \left(\frac{\partial \psi_{1,2}}{\partial z}\frac{\partial}{\partial x} - \frac{\partial \psi_{1,2}}{\partial x}\frac{\partial}{\partial z}\right)\left(\frac{\partial^2}{\partial x^2} + \frac{\partial^2}{\partial y^2}\right)\psi_{1,2} = 0. \tag{3.85}$$

The boundary conditions now read

$$\psi_{1,2} = \psi, \tag{3.86a}$$

$$\delta \frac{\partial\psi_{1,2}}{\partial z} = \frac{\partial\psi}{\partial z}, \quad z = \pm 1 + \varepsilon\zeta. \tag{3.86b}$$

As it is easily seen, the equations describing the waves outside of the middle layer do not belong to type (3.30). This situation needs a more detailed discussion for the zeroth approximation when nonlinearity and dispersion and absent ($\varepsilon = 0$, $\delta = 0$, respectively). Here an interesting quality must be underlined: the equations describing the waves in the middle layer at $\varepsilon = \delta = 0$

$$\frac{\partial\rho}{\partial t} + \frac{d\rho_0}{dz}\frac{\partial\psi}{\partial x} = 0, \tag{3.87a}$$

$$\frac{\partial}{\partial z}\left(\rho_0 \frac{\partial^2\psi}{\partial z\partial t}\right) + \frac{\partial\rho}{\partial x} = 0, \tag{3.87b}$$

together with the boundary conditions

$$\frac{\partial\psi}{\partial z} = 0, \quad z = \pm 1 \tag{3.88}$$

constitute a closed system and are not coupled with the equations describing the wave motion in the homogeneous layers. This means physically that the wave motion at this approximation coincides with the motion in a channel with solid covers at $z = \pm 1$. In addition, the nondispersive character of these waves is obvious. The solution itself is given by

$$\psi = M(z)v(x - ct), \tag{3.89a}$$

$$\rho = \frac{1}{c}\frac{d\rho_0}{dz} M(z)v(x - ct), \tag{3.89b}$$

where $M(z)$ satisfies the boundary problem

$$\frac{d}{dz}\left(\rho_0 \frac{dM}{dz}\right) - \frac{1}{c^2}\frac{d\rho_0}{dz} M = 0, \tag{3.90a}$$

$$\frac{dM}{dz} = 0, \quad z = \pm 1. \tag{3.90b}$$

Thus an interesting situation exists here: despite the "nonstandard" equations describing the motion in homogeneous layers, their solutions will be "switched on" to the solutions of the stratified layer only by the small dispersion parameter. This situation shows explicitly the possibility to construct evolution equations for the long-wave approximation. The asymptotic procedure for such a case was proposed earlier [57, 77] and we shall repeat it here briefly.

The solution to the system (3.83)-(3.86) in the case of small but finite ε and δ is sought in the form of the asymptotic series

$$\psi = M(z)v(\xi,\tau) + \sum_{i,j} \varepsilon^i \delta^j \psi^{ij}(z,\xi,\tau), \tag{3.91}$$

$$\rho = \frac{1}{c}\frac{d\rho_0}{dz} M(z)v(\xi,\tau) + \sum_{i,j} \varepsilon^i \delta^j \rho^{ij}(z,\xi,\tau), \tag{3.92}$$

and

$$\frac{\partial v}{\partial \tau} + \sum_{i,j} \varepsilon^{i-1}\delta^j \Phi^{ij} \{v,\xi,\tau\} = 0, \tag{3.93}$$

where $\xi = x - ct$. Actually we could use only the one-parameter ($\varepsilon \sim \delta$) expansion as in Section 3.2, but using the multi-parameter expansions leads to less complicated algebra and is therefore preferable. We restrict ourselves to the first approximation, because our main aim is to demonstrate the essential differences in comparison with the problems discussed earlier.

The easiest task is associated with the nonlinearity, and at the first approximation with respect to ε, together with $\delta = 0$, the boundary conditions at interface take the form

$$\frac{\partial \psi}{\partial z} + \varepsilon\xi \frac{\partial^2 \psi}{\partial z^2} = 0, \quad z = \pm 1 \tag{3.94}$$

and do not contain any function describing motion in other unstratified layers. Comparing this problem with the previous one at the zeroth approximation (with the "solid covers") discussed above, we conclude that only the boundary conditions are different. Physically this means that the boundaries are no longer rigid. This is not the principal difference, and actually the explicit change is seen only in the coefficient Φ^{10} of the nonlinear term in the evolution equation (3.73)

$$\Phi^{10} = \bar{\alpha} v \frac{\partial v}{\partial \xi}, \tag{3.95a}$$

$$\bar{\alpha} = \frac{\frac{1}{2}\int_{-1}^{1} M^3 \frac{d^2\rho_0}{dx^2} + M(1)\frac{d\rho_0(1)}{dz} + M(-1)\frac{d\rho_0(-1)}{dz}}{\int_{-1}^{1} M^2 \frac{d\rho_0}{dz}\, dz}. \tag{3.95b}$$

Next let us discuss the first approximation with respect to the dispersion parameter δ assuming $\varepsilon = 0$. The corrections ρ^{01} and ψ^{01} are determined from the equations

$$-c\frac{d\rho^{01}}{d\xi} + \frac{d\rho_0}{dz}\frac{\partial\psi^{01}}{\partial\xi} - \frac{1}{c}\frac{d\rho_0}{dz} M_0\Phi^{01}, \tag{3.96}$$

$$-c\frac{\partial}{\partial z}\left(\rho_0\frac{\partial^2\psi^{01}}{\partial z\partial\xi}\right) + \frac{\partial\rho^{01}}{\partial\xi} = \frac{d}{dz}\left(\rho_0\frac{dM}{dz}\right)\Phi^{01}, \tag{3.97}$$

subject to the boundary conditions

$$\frac{\partial\psi^{01}}{\partial z} = \frac{\partial\psi_{\pm}^{00}}{\partial z},\quad z = \pm 1. \tag{3.98}$$

As is easily seen, the corrections are generated by the right-hand side $\psi_{\pm}^{00}$. Let us determine the correction for the upper layer, for example. At the zeroth approximation ψ_{+}^{00} satisfies the Laplace equation

$$\frac{\partial}{\partial t}\left(\frac{\partial^2}{\partial x^2} + \frac{\partial^2}{\partial z^2}\right)\psi_{+}^{00} = 0, \tag{3.99}$$

and the conditions

$$\psi_{+}^{00} = \psi^{00} = M(1)v(\xi,\tau),\quad z = \pm 1, \tag{3.100a}$$

$$\psi_+^{00} = 0, \quad z \to \infty. \tag{3.100b}$$

The solution to (3.99) together with (3.100) is the harmonic function

$$\psi_+^{00} = -\frac{M(1)}{\pi} \int_{-\infty}^{+\infty} \frac{(z-1)v(\xi',\tau)d\xi'}{(\xi'-\xi)^2 + (z-1)^2}, \tag{3.101}$$

where the integration is understood in the value principal sense. The boundary condition (3.98) needed for the system (3.96), (3.97) now takes the form

$$\frac{\partial\psi^{01}}{\partial z} = \frac{M(1)}{\pi} \int_{-\infty}^{\infty} \frac{v(\xi',\tau)d\xi'}{\xi - \xi'} \tag{3.102}$$

for z = 1. The condition for z = -1 is the same with M(1) replaced by M(-1). Equation (3.102) yields

$$\psi^{01} = \frac{R(z)}{\pi} \frac{\partial}{\partial\xi} \int_{-\infty}^{\infty} \frac{v(\xi',\tau)d\xi'}{\xi - \xi'}, \tag{3.103}$$

and, consequently, the equations (3.96), (3.97)

$$\Phi^{01} = \frac{P}{\pi} \frac{\partial^2}{\partial\xi^2} \int_{-\infty}^{\infty} \frac{v(\xi',\tau)d\xi'}{\xi - \xi'}, \tag{3.104}$$

$$\rho^{01} = \frac{\Theta(z)}{\pi} \frac{\partial}{\partial\xi} \int_{-\infty}^{\infty} \frac{v(\xi',\tau)d\xi'}{\xi - \xi'}. \tag{3.105}$$

Here the constant P and the functions R(z) and $\Theta(z)$ must be found from the boundary problem

$$-c\Theta + \frac{d\rho_0}{dz} R = \frac{1}{c} \frac{d\rho_0}{dz} MP, \tag{3.106a}$$

$$-c \frac{d}{dz}\left(\rho_0 \frac{dR}{dz}\right) + \Theta = \frac{d}{dz}\left(\rho_0 \frac{dM}{dz}\right) P. \tag{3.106b}$$

This system is equivalent to the single equation

$$-c^2 \frac{d}{dz}\left(\rho_0 \frac{dR}{dz} + \frac{d\rho_0}{dz} R\right) = \frac{2P}{c} \frac{d\rho_0}{dz} M \tag{3.107}$$

with the conditions

$$\frac{dR(\pm 1)}{dz} = M(\pm 1). \tag{3.108}$$

Since the operator on the left-hand side of (3.107) is self-conjugate, the orthogonality conditions are obtained by multiplying equation (3.107) by M and integrating them over z from -1 to +1, taking condition (3.108) into account. This action first permits us to determine

$$P = -\frac{c^3}{2} \frac{\rho_0(1)M^2(1) + \rho_0(-1)M^2(-1)}{\int_{-1}^{1} \frac{d\rho_0}{dz} M^2 dz}, \tag{3.109}$$

and then, consequently, the functional Φ^{01}. Now it is not difficult to deduce the evolution equation of the first approximation

$$\frac{\partial v}{\partial \tau} + \varepsilon\bar{\alpha} v \frac{\partial v}{\partial \xi} + \frac{\delta}{\pi} P \frac{\partial^2}{\partial \xi^2} \int_{-\infty}^{+\infty} \frac{v(\xi', \tau)}{\xi - \xi'} d\xi'. \tag{3.110}$$

This is called the Benjamin-Ono equation [77], and it differs from the Korteweg-de Vries equation in its last term. This term, modelling integral dispersion, is created by wave motion in the unstratified layers of infinite depth. If the depth is large but finite then the resulting evolution equation may be written in a generalized form [97, 116] with the limit cases either the Korteweg-de Vries equation for $\lambda \gg H$ or the Benjamin-Ono equation (3.110) for $H \gg \lambda \gg h$.

Finally we wish to emphasize that in complicated situations when the modes depend on the wave number and the equations are not transformed to standard form, the search for an optimal asymptotic procedure depends essentially on the knowledge and practice of the researcher. The most cumbersome approach is certainly the spectral method.

4 Applications. Simple evolution equations

In this Chapter the well-known simple evolution equations governing the nonlinear wave propagation in nondispersive media are discussed briefly. The emphasis is placed on the physical analysis. This part has been written from the viewpoint of a graduate student in mathematical physics and may be used as a reference-book for the analysis of concrete wave processes.

4.1 The equation of a simple wave

Riemann wave. The equation of a simple wave has the following form

$$\frac{\partial u}{\partial s} + \alpha u \frac{\partial u}{\partial \xi} + Q(s)u = 0. \tag{4.1}$$

Here the wave variable u is normalized by its maximum value and the independent variables s and ξ are normalized by the characteristic scale of the nonlinearity and by the wavelength, respectively. The parameter α is dimensionless. By means of the transformation

$$v = u \{\exp \int Qds\}, \tag{4.2a}$$

$$z = \int \alpha \exp \{- \int Qds\}ds \tag{4.2b}$$

the equation (4.1) is reduced to the homogeneous equation of a simple wave

$$\frac{\partial v}{\partial z} + v \frac{\partial v}{\partial \xi} = 0. \tag{4.3}$$

The detailed analysis of this equation has been presented by many authors [46, 54, 96, 97, 122]. The solution to equation (4.3) subject to the boundary condition $v(z = 0) = v^0(z)$ may be written implicitly in the form of a simple (Riemann) wave

$$v = v^0(\xi - zv), \tag{4.4}$$

which describes the nonlinear steepening of the wave profile and governs all the distortions of the smooth perturbation $v^0(\xi)$ given at $z = 0$. If f denotes the inverse function to v^0 then the simple wave may also be written as

$$\xi = zv + f(v). \tag{4.5}$$

Here the term zv describes the distortion of the initial profile f(v). The nonlinear distortions accumulate with passage of z and the wave profile steepens. Finally at a certain $z = z_S$ a discontinuity forms at the point with the maximum initial steepness $dv^0/d\xi$. The coordinate z_s, determined by

$$z_s = \left\{ \left| \frac{dv^0}{d\xi} \right|_{max} \right\}^{-1}, \tag{4.6}$$

is obtained by using the method of characteristics. The solution (4.5) is correct for the full profile including the discontinuity. Therefore expression (4.5) may also be applied to the points of discontinuity $v^{\pm}$

$$\xi = zv^+ + f(v^+), \tag{4.7a}$$

$$\xi = zv^- + f(v^-), \tag{4.7b}$$

where v^+ indicates the value behind the discontinuity and v^- the value in front of it. When differentiating one of the equations (4.7), say (4.7a), we obtain after certain transformations [78]

$$\frac{v^+ - v^-}{2} \frac{dz}{dv^+} + z = \frac{df(v^+)}{dv^+}. \tag{4.8}$$

Combining equations (4.7) we arrive at the additional equation

$$z(v^+ - v^-) = f(v^+) - f(v^-). \tag{4.9}$$

The system of equations (4.8), (4.9) governs the variables $v^{\pm}$ which determine the shock amplitude. At certain boundary conditions $f(v)$ the solution of (4.5), (4.8) and (4.9) may be obtained in an explicit form. Thus, for example, for an initial sine-type excitation the distortion up to shock-wave formation is described by the Fubini solution, and the further N-wave distortion at large time - by the Fay solution (details given in [9, 19, 27, 37, 97]). If the initial exictation is of the arbitrary form, then the explicit solution may be found by means of a series expansion in the vicinity of the discontinuity.

Discontinuous solutions. The solutions of the equations (4.8), (4.9) may be constructed explicitly for several types of boundary conditions. Let us start with a bounded pulse where the discontinuity has formed at the point with zero amplitude. In this case the variable v^+ is associated with the value $v^- = 0$. Consequently, equation (4.8) has the simpler form

$$\frac{v^+}{2}\frac{dz}{dv^+} + z = \frac{df}{dv^+} . \tag{4.10}$$

Integrating this equation by parts we find

$$z = \frac{2}{(v^+)^2}\int_{v^+} \zeta \frac{df}{d\zeta}\, d\zeta. \tag{4.11}$$

The integration is to be performed from zero to v^+ when the discontinuity propagates with the wave front, and from 1 to v^+ when the discontinuity has reached the maximum amplitude of the wave profile (here and later we assume that the wave amplitude is normalized so that its maximum value is one). As an example, let us find the integral (4.11) for the pulse $v = (1 - \xi)^{1/m}$ or $f = 1 - v^m$, where $m > 0$. If $m = 1$ the pulse is of the triangular form. Substituting df/dv into the integral (4.11) and performing the integration from zero to v^+ we get

$$(v^+)^{m+1} + \frac{m + 1}{2m} z(v^+)^2 - 1 = 0. \tag{4.12}$$

For a wave of the triangular form ($m = 1$) the well-known expression $v^+ = (1 + z)^{-1/2}$ follows easily. If $m = 3$ then an explicit expression for $v^+(z)$ is also obtained without any difficulties. In Table 4.1 the results

Pulse shape at $z = 0$	Amplitude $v^+(z)$	Length $\Theta_s(z)$	Shock formation coordinate
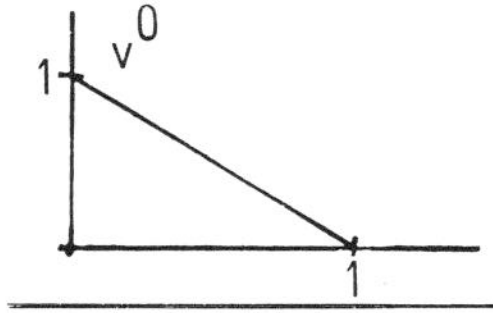 $v^0 = 1 - \zeta$ $0 \leqq \zeta \leqq 1$	$v^+ = (1 + z)^{-1/2}$	$\Theta_s = (1 + z)^{-1/2}$	$z_s = 0$
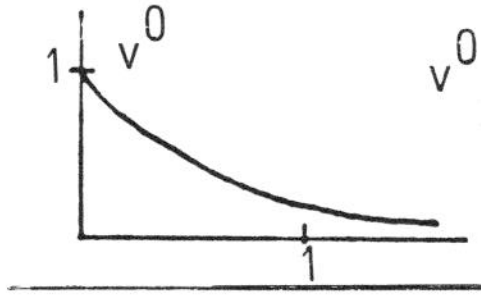 $v^0 = \exp(-\zeta)$ $\zeta \geqq 0$	$v^+ = \dfrac{2}{1 + (1 + 2z)^{1/2}}$	$\Theta_s = \dfrac{1 + (e-1)(1+2z)^{1/2}}{e}$	$z_s = 0$
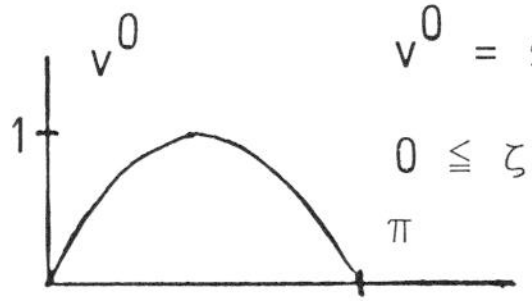 $v^0 = \sin \zeta$ $0 \leqq \zeta \leqq \pi$	$v^+ = 2\,\dfrac{(z - 1)^{1/2}}{z}$ $z \geqq 1$	$\Theta_s = \pi + 2[(z-1)^{1/2} +$ $+ \text{arctg}\,(z-1)^{1/2}],$ $z \geqq 1$	$z_s = 1$
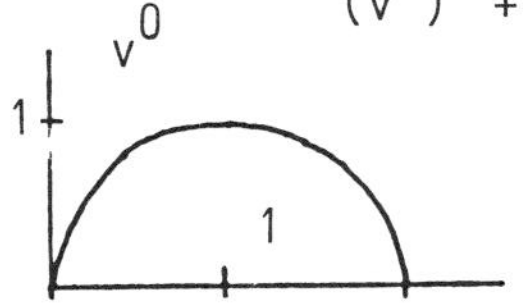 $(v^0)^2 + (\zeta-1)^2 = 1$ $0 \leqq \zeta \leqq 2$	$z = \dfrac{\arcsin v^+}{(v^+)^2} - \dfrac{(1-(v^+)^2)^{1/2}}{v^+}$ $z \to \infty \quad v^+ = \sqrt{\pi/z}$	$z \to \infty$ $\Theta_s = \sqrt{\pi z}$	$z_s = 0$

Table 4.1. The parameters of various shock pulses

are presented for several types of pulses frequently used in practice [30, 33, 87].

At large distances ($z >> 1$) the integral in the expression (4.11) is actually the area of the initial excitation. It means that the solution tends to the only possible asymptotic solution

$$v^+ \sim z^{-1/2}. \tag{4.13}$$

Another solution may be constructed for a symmetrical discontinuity formed at the point where $v = 0$ and connecting two values v^+ and $v^- = -v^+$. Using equation (4.9) we find

$$v^+ = \frac{1}{z} f(v^+). \tag{4.14}$$

Let us discuss this solution more closely. The discontinuity developing at the part of the harmonic wave decays at large distances according to (4.14) like $v^+ \sim z^{-1}$. This is the well-known asymptotic dependence for N-waves. In the case of a symmetrical unperiodic wave the discontinuity decays differently.

Firstly, the function $f(v^+ \to 0)$ may increase with no limit and secondly, the asymptotic behaviour $v^+(z)$ will be determined by the form of the "tail" portion. For example, if $v(\xi)$ decays as ξ^{-m}, then the amplitude of the discontinuity is governed by

$$v^+ \sim z^{-m/m+1}. \tag{4.15}$$

If the pulse has finite energy, then the amplitude of the discontinuity decays more slowly than the amplitude of the N-wave. Thus, the solution of the one-polar pulse has a single asymptotic determined by its area, while the solutions for pulses with zero area have no such a single asymptotic.

All the solutions obtained above concern a homogeneous medium. By means of the transformation (4.2) it is possible to deduce solutions in inhomogeneous media on the basis of the solutions to the homogeneous equation (4.3).

The transfer integral for an inhomogeneous medium. Expressions (4.2) determine all the peculiarities of simple and shock waves in an inhomogeneous medium. In order to analyze integral (4.2b) we use the dependence of Q on the ray coordinate s. In acoustics, for example, it has the form

$$Q = \frac{c\rho}{2S} \frac{d}{ds} \left(\frac{S}{c\rho}\right). \tag{4.16}$$

Substituting (4.16) into (4.2b) and returning to a dimensional ray coordinate, we have

$$z = \frac{1}{R_*} \int_{s_0}^{s} \frac{q(s')}{q_0} \, ds', \tag{4.17a}$$

$$q = \alpha (Sc^5 \rho)^{-1/2}, \tag{4.17b}$$

$$R_* = \lambda_0/(\alpha_0 M_0). \tag{4.17c}$$

Here λ_0, M_0 and α_0 are the wavelength, Mach number and the parameter of nonlinearity, respectively, in the vicinity of the source at the point with the ray coordinate s_0; R_* is the scale of the nonlinearity. Several particular cases of integral (4.17) have been discussed earlier by many authors [25, 29, 65, 69, 70, 71, 97, 113]. Here we shall present some physically important examples.

Example 4.1. Cylindrical and spherical waves in a homogeneous medium. The transfer integral (4.17a) now has the form

$$z = \frac{1}{R_*} \int_{s_0}^{s} \frac{ds'}{A^{1/2}} , \tag{4.18}$$

where A is the dimensionless cross-section of the ray tube, $A = (s/s_0)^n$, $n = 1,2$ for cylindrical and spherical waves, respectively. After simple calculations we obtain

$$z = \begin{cases} \dfrac{s_0^{1/2}}{R_*}(s^{1/2} - s_0^{1/2}), & n = 1, \quad (4.19a) \\ \\ \dfrac{s_0}{R_*}\ln\dfrac{s}{s_0}, & n = 2. \quad (4.19b) \end{cases}$$

Obviously divergent waves always tend to shock formation due to the nonlinear profile distortions. This is also correct for convergent spherical wave. In the case of a convergent cylindrical wave, shock formation occurs at a distance

$$s_s = s_0\,(1 - z_s R_*/s_0)^2. \tag{4.20}$$

Therefore, the shock formation condition yielded by this expression, is

$$z_s \geqq s_0/R_*. \tag{4.21}$$

Thus, for the shock formation in a convergent cylindrical wave, either a rather long distance s_0 is needed, or a small characteristic scale of non-linearity R_*.

Example 4.2. Cylindrical and spherical waves in cylindrically and spherically layered media. Here we shall discuss the case of a power law dependence. Such a situation occurs in stellar atmospheres [123]. Here we have

$$c = c\,(s/s_0)^k, \tag{4.22a}$$

$$\rho = \rho_0(s/s_0)^{\ell}, \tag{4.22b}$$

$$\alpha = \alpha_0(s/s_0)^m, \tag{4.22c}$$

$$S = S_0(s/s_0)^n. \tag{4.22d}$$

Substituting (4.22) into (4.17) we arrive at the integral

$$z = \frac{1}{R_*} \int_{s_0}^{s} (s_0/s)^{\delta} \, ds, \tag{4.23}$$

where $\delta = m - 5k/2 - \ell/2 - n/2$ and, as previously, $n = 1,2$ for cylindrical and spherical waves, respectively. Further calculation gives

$$z = \begin{cases} \dfrac{s_0}{(1+\delta)R_*} [(s/s_0)^{1+\delta} - 1], \quad \delta \neq 1 & (4.24a) \\[2ex] \dfrac{s_0}{R_*} \ln(s/s_0), \quad \delta = 1. & (4.24b) \end{cases}$$

The next question is concerned with possible shock formation. If the parameter $\delta > -1$, then the discontinuity always forms in a divergent wave. In case of an isothermal medium ($k = 0$) with constant nonlinearity ($m = 0$) this condition may be replaced by $\ell + n < 2$, i.e. we have

$$\ell < 2 - n. \tag{4.25}$$

Thus a discontinuity in a cylindrical wave may form even in case of a wave propagating into a medium with increasing density ($\ell < 1$).

Example 4.3. Plane waves in an exponentially inhomogeneous medium. We assume here that the density varies exponentially like

$$\rho = \rho_0 \exp(-s/H), \tag{4.26}$$

which is characteristic of the lower layers of the Earth's atmosphere. The variable z, calculated from (4.17) now becomes

$$z = \begin{cases} \dfrac{2H}{R_*} [\exp(-s/2H) - 1], & (4.27a) \\[2ex] \dfrac{2H}{R_*} [1 - \exp(-s/2H)] & (4.27b) \end{cases}$$

for the waves propagating in the direction of decreasing and increasing densities, respectively.

According to condition (4.6), for a sine wave we have $z_s = 1$, and therefore a shock wave will be formed at a distance

$$s_s = 2H \ln (1 \pm R_*/2H). \qquad (4.28)$$

The shock wave always forms when the wave propagates in the direction of a density decrease and there exists an asymptotic decay law [90, 95]. If the wave propagates in the direction of a density increase, the shock wave will form only in case of rather high intensity when the condition

$$R_* < 2H \qquad (4.29)$$

is satisfied.

Example 4.4. Critical gradients. Let us formulate the shock formation condition for a wave propagating in an inhomogeneous medium. The shock formation coordinate is determined by the steepness of the initial excitation. As the condition $z \geqq z_s$ must always be satisfied, the initial steepness, according to the expression (4.6), must satisfy the condition

$$g = |dv^0/d\xi|_{max} \geqq 1/z. \qquad (4.30)$$

If the homogeneity has such a character that the integral (4.25) gives an increasing function z(s), then the condition (4.30) is satisfied beyond a certain distance s_s. If, however, integral (4.2b) leads to an asymptotic value (i.e. $z(s \to \infty) \to z_\infty$ = const.) then the shock formation condition

$$g \geqq z_\infty^{-1}$$

can only be satisfied for initial excitations which have steepness higher than a critical value. The N-waves with a critical gradient propagate without any changes in their form [67, 87]. There may also exist a possibility that the gradient g is higher than the critical one and that a shock wave has already been formed. The wave will be developing and tending to its asymptotic law of decay only when the condition $g \gg z^{-1}$ is satisfied. If this condition is not satisfied then the discontinuity will develop until a certain profile configuration at large $s \gg s_0$, and afterwards will propagate as a wave with a constant profile.

Example 4.5. High-frequency processes in relaxing media. The exact formulation of the evolution equations for such processes will be given in Section 4.3. If the high-frequency approximation is possible then the evolution equation reduces to the simple wave equation (4.1) [34, 56, 65]. For example, for media with the high thermal conductivity, expressions (4.2) read

$$v = u/u_0 \exp(s/H), \tag{4.31a}$$

$$z = \frac{H}{R_*} (1 - \exp(-s/H)), \tag{4.31b}$$

$$H = 2\gamma\chi/(\gamma^{-1})c_T, \tag{4.31c}$$

where χ is the coefficient of the thermal conductivity, c_T is the isothermal sound velocity, H is the characteristic scale of the thermal relaxation. The shock formation coordinate s_s is now determined by

$$s_s = -H \ln (1 - z_s R_*/H). \tag{4.32}$$

If the nonlinearity is high (i.e. $R_* \ll H$), then the shock formation coordinate s_s does not depend on the scale of relaxation. The nonlinearity decrease makes the shock formation coordinate s_s increase. This increase goes faster than in a nonrelaxing medium and finally, when $R_* = H/z_s$, a shock will not form at all and the profile remains smooth.

4.2 The Burgers equation

Waves in media with losses. Waves in inhomogeneous media with losses are governed by the modified Burgers equation

$$\frac{\partial u}{\partial s} + \alpha u \frac{\partial u}{\partial \xi} + Q(s)u - b \frac{\partial^2 u}{\partial \xi^2} = 0. \tag{4.33}$$

Here b is the dissipation coefficient. Introducing the transformation (4.2), equation (4.33) is reduced to the usual Burgers equation with a variable coefficient describing the losses

$$\frac{\partial v}{\partial z} + v\frac{\partial v}{\partial \xi} - \Gamma^{-1}(z)\frac{\partial^2 v}{\partial \xi^2} = 0. \qquad (4.34)$$

The variable Reynolds number Γ characterizes the scaled relationship of the nonlinear and dissipative effects together with the effects of the inhomogeneity

$$\Gamma = \frac{\alpha}{b}\exp(-\int Q ds). \qquad (4.35)$$

According to (4.2), the function $\Gamma(z)$ is determined by

$$\Gamma(z) = \left\{b\frac{ds}{dz}\right\}^{-1}. \qquad (4.36)$$

It is easy to show that for spherical waves $\Gamma^{-1} \sim \exp z$, for cylindrical waves $\Gamma^{-1} \sim z$, and for plane waves propagating in the direction of density decrease in an isothermal medium $\Gamma^{-1} \sim z$. It should be pointed out that in an inhomogeneous medium even with a comparatively small dissipation coefficient b, the cross-section of the ray-tube may increase so much that the coefficient Γ^{-1} will also increase. This means that equation (4.34) may be replaced at such distances from the source by the linear parabolic equation

$$\frac{\partial v}{\partial z} = \Gamma^{-1}\frac{\partial^2 v}{\partial \xi^2}. \qquad (4.37)$$

To the contrary, the effective Reynolds number Γ may increase when the wave propagates in a strongly dissipative medium (the ray-tube may be, for example, compressed in the course of time). Equation (4.34) then reduces to the equation of simple waves (4.3). In both limit cases the solutions are known. For the special cases of function $\Gamma(z)$, including spherical and cylindrical waves in dissipative media, the approximate solutions of the equation (4.34) are known [20, 21, 69-71].

It is of interest to note that wave propagation in an inhomogeneous dissipative medium may be governed by the homogeneous equation (4.34). This is possible when the nonlinear parameter α changes along the ray coordinate in a certain way depending on the inhomogeneity. In this case

$$\alpha(s) = \text{const}\cdot\exp\left(\int Q(s)ds\right). \tag{4.38}$$

In acoustics, according to the expression (4.16), the Reynolds number is constant provided

$$\frac{1}{\alpha}\left(\frac{S}{c\rho}\right)^{1/2} = \text{const.} \tag{4.39}$$

is satisfied. In particular, when a plane wave propagates along the gradient of an inhomogeneity, the dissipation coefficient remains constant at the constant impedance $I = \rho c$.

There exists a vast amount of literature dealing with the Burgers equation and its solution [8, 19, 36, 49, 67, 97, 101, 122]. The Burgers equation is one of the rare nonlinear equations having a closed exact solution. This is found by means of the Cole-Hopf transformation [15, 42] which will be presented for a wide class of media with constant impedance.

The Cole-Hopf transformation. Setting

$$v = -2\Gamma^{-1}\frac{1}{\varphi}\frac{\partial\varphi}{\partial\xi}, \tag{4.40}$$

it is possible to reduce equation (4.34) to the diffusion equation

$$\frac{\partial\varphi}{\partial z} = \Gamma^{-1}\frac{\partial^2\varphi}{\partial\xi^2}. \tag{4.41}$$

The initial condition for (4.34) $v = v^0\ (z = 0,\xi)$ is transformed respectively into

$$\varphi(0,\ ) = \exp\left\{-\frac{\Gamma}{2}\int_0^{\xi} v^0(\xi')d\xi'\right\}. \tag{4.42}$$

The solution to equation (4.41) subject to condition (4.42) is

$$\varphi(z,\xi) = \frac{1}{2(\pi z\Gamma^{-1})^{1/2}}\int_{-\infty}^{+\infty}\exp\left\{-\frac{(\xi-\xi')^2}{2(z\Gamma^{-1})^{1/2}} + \frac{\Gamma}{2}\int_0^{\xi_1} v^0(\xi)d\xi\right\}d\xi'. \tag{4.43}$$

Two characteristic examples are now presented. First, we consider the propagation of a bounded ($0 \leqq \xi \leqq \xi_A$) pulse

$$v^0(\xi) = \kappa(\xi)[H(\xi) - H(\xi - \xi_A)] \tag{4.44}$$

where $H(\xi)$ is the Heaviside step-function. The final result is

$$\varphi(\xi,z) = \frac{1}{2}\,\mathrm{erfc}\,[\tfrac{1}{2}\xi(z\Gamma^{-1})^{-1/2}] + \\ + \frac{1}{2}\,\mathrm{erfc}\,[\tfrac{1}{2}\,(\xi - \xi_A)(z\Gamma^{-1})^{-1/2}] + \\ + \frac{1}{2(\pi z\Gamma^{-1})^{1/2}} \int_0^{\xi_A} \exp\,[B(\xi,\xi')]d\xi', \tag{4.45a}$$

$$B(\xi,\xi') = -\frac{(\xi - \xi')^2}{2(z\Gamma^{-1})^{1/2}} + \frac{\Gamma}{2}\int_0^{\xi'} v^0(\xi)d\xi, \tag{4.45b}$$

$$\mathrm{erfc}(x) = 1 - \Phi(x), \tag{4.45c}$$

$$\Phi(x) = \frac{2}{\pi^{1/2}} \int_0^x \exp(-t^2)dt. \tag{4.45d}$$

Second, a pulse of the bell-shape form

$$v^0(\xi) = \xi(1 + q^2\xi^2)^{-2}, \quad q = \mathrm{const} \tag{4.46}$$

is considered. Here the final result is

$$v(\xi,z) = v_0(\xi) \left[1 + \frac{1}{2}\,\mathrm{erfc}(-\frac{1}{2}\,\xi\,\frac{\exp\,f(\xi)}{(z\Gamma^{-1})^{1/2}}\right]^{-1}, \tag{4.47a}$$

$$f(\xi) = \frac{1}{4}\,\Gamma q^{-2}\,[1 - q^2\xi^2)^{-1}]. \tag{4.47b}$$

Obviously at $z \to 0$ this expression describes the shock profile.

4.3 An integro-differential equation

The one-dimensional integro-differential equations

$$\frac{\partial v}{\partial z} + v\frac{\partial v}{\partial \xi} + \frac{\partial}{\partial \xi}\int_0^{\xi} \frac{\partial v}{\partial \xi'} K(\xi' - \xi)d\xi' = 0 \qquad (4.48)$$

belong to the most general class of the evolution equations governing the wave distortions in media with different dissipation and dispersion. The kernel function $K(\xi)$ and the complex dispersion function $k(\omega)$ are related through

$$k(\omega) = \omega\int_{-}^{+} K(\xi)\exp(-i\omega\xi)d\xi, \qquad (4.49a)$$

$$K(\xi) = \frac{1}{2\pi}\int_{-\infty}^{+\infty} \frac{k(\omega)}{\omega}\exp(i\omega\xi)d\xi. \qquad (4.49b)$$

The kernel function $K(\xi)$ is usually determined from the constitutive equation. Thus, for example, the kernel function for a standard viscoelastic medium is an exponential one, and the corresponding evolution equation has the form

$$\frac{\partial v}{\partial z} + v\frac{\partial v}{\partial \xi} - m\frac{\partial}{\partial \xi}\int_0^{\xi} \frac{\partial v}{\partial \xi'}\exp\left\{-\frac{\xi - \xi'}{\Theta}\right\}d\xi' = 0, \qquad (4.50)$$

where m and Θ are certain constant parameters (for details see [24]). The exponential kernel is typical for many relaxation processes [24, 97]. It concerns media with chemical reactions [6, 74] and the salt solutions as, for example, in ocean water [38]. If the relaxation is caused by the thermal processes, then beside the exponential kernel, a square root-type kernel must also be taken into account ($K \sim |\xi - \xi'|^{1/2}$) [50, 56, 58]. In the limit cases $\Theta \gg 1$ (the high-frequency processes) and $\Theta \ll 1$ (the low-frequency processes) the evolution equation (4.50) has asymptotic expansions and is reduced to certain differential equations [24, 45, 97]. In the general case equation (4.50) may be reduced to a differential equation of higher order

$$(\Theta\frac{\partial}{\partial \xi} + 1)\left(\frac{\partial v}{\partial z} + v\frac{\partial v}{\partial \xi}\right) = m\Theta\frac{\partial^2 v}{\partial \xi^2}. \qquad (4.51)$$

<u>Example 4.6.</u> <u>The stationary wave.</u> The solution to the equation (4.51) is sought in the form of a stationary wave

$$v(z,\xi) = v(\zeta), \quad \zeta = \xi - \kappa z, \quad \kappa = \text{const.} \tag{4.52}$$

Substituting (4.52) into (4.51) we obtain after integration

$$\frac{dv}{d\zeta} = -\frac{1}{2\Theta} \frac{(v - v_1)(v - v_2)}{(v - \kappa_0)}, \tag{4.53}$$

$$\kappa_0 = \kappa + m \tag{4.54a}$$

$$v_{1,2} = \kappa \pm (\kappa^2 + 8A)^{1/2}, \tag{4.54b}$$

where A is the integration constant. The dependence of the stationary shock wave profile on the relaxation amplitude m is clearly seen from equation (4.53). The profile itself is unsymmetric and may be expressed implicitly according to (4.53) as

$$\zeta = -2\Theta \ln (v - v_1)^{\alpha_1}(v - v_2)^{\alpha_2}, \tag{4.55}$$

$$\alpha_1 = \frac{v_1 - \kappa_0}{v_1 + v_2}, \tag{4.56a}$$

$$\alpha_2 = \frac{v_2 - \kappa_0}{v_1 + v_2}. \tag{4.56b}$$

A stationary wave can exist provided $\kappa_0 > v_1$. If the relaxation amplitude m decreases then the value of κ_0 may be equal to or even less than the maximum field variable v_1. The solution (4.55) is then non-unique and a stationary wave can no longer exist. Physically this means that the relaxation is weak and cannot compete with the nonlinear distortions.

4.4 The equation of nonlinear rays

Let us consider the influence of nonlinearity on the distortion of ray trajectories. A rather good example here is shock wave propagation with a triangular profile. We start from the equations (2.53) and consider the two-dimensional motion. The equations (2.53a), (2.53b) are reduced to

$$\frac{\partial A}{\partial s} = \frac{\partial \varphi}{\partial m}, \tag{4.57a}$$

$$\frac{\partial\varphi}{\partial s} = -\frac{\alpha}{2A}\frac{\partial M}{\partial m} . \tag{4.57b}$$

Here $A = S/S_0$ is the dimensionless cross-section area of the ray tube, M is the Mach number at the front of the triangular pulse, φ is the deviation angle of the ray from its initial trajectory, s and m are the dimensionless longitudinal and the transverse coordinates, respectively. Equation (2.53d) is easily solved for a wave of triangular profile and its solution is [31]

$$\frac{\partial}{\partial s}\frac{1}{AM^2} = \frac{\alpha}{2\Pi A^{1/2}} , \tag{4.58}$$

where $\Pi = \lambda(m)M_0(m)/2$ is the initial area of the triangular pulse and λ is its length.

The solution of the system (4.57), (4.58) is sought by the perturbation method assuming the nonlinear parameter α is a small parameter. Let us assume

$$A = A_0 + \alpha A_1, \tag{4.59a}$$

$$M = M_0 + \alpha M_1, \tag{4.59b}$$

$$\varphi = \varphi_0 + \alpha\varphi_1$$

and restrict ourselves to first order corrections. The linear approximation is determined from the system

$$\frac{\partial A_0}{\partial s} = \frac{\partial\varphi_0}{\partial m} , \tag{4.60a}$$

$$\frac{\partial\varphi_0}{\partial s} = 0, \tag{4.60b}$$

$$\frac{\partial}{\partial s}\left(\frac{1}{A_0 M_0^2}\right) = 0. \tag{4.60c}$$

For cylindrical and plane waves it yields

$$M_0 = \bar{M}_0(m)/S^{1/2}, \tag{4.61a}$$

$$A_0 = S, \tag{4.61b}$$

$$\varphi = \varphi_0^{(0)}(m); \tag{4.61c}$$

$$M_0 = \bar{M}_0(m), \tag{4.62a}$$

$$A_0 = 1, \tag{4.62b}$$

$$\varphi_0 = \text{const}, \tag{4.62c}$$

respectively. The first order approximation is governed by the system (the index 1 is omitted)

$$\frac{\partial A}{\partial s} = \frac{\partial \varphi}{\partial m}, \tag{4.63a}$$

$$\frac{\partial \varphi}{\partial s} = -\frac{1}{2A_0}\frac{\partial M_0}{\partial m}, \tag{4.63b}$$

$$\frac{\partial}{\partial s}\left(\frac{A}{A_0} + 2\frac{M}{M_0}\right) = -\frac{a(m)}{A_0^{1/2}}. \tag{4.63c}$$

Here $z(m) = A_0 M_0^2/2\Pi(m)$. System (4.63) can easily be integrated. The corresponding solution for cylindrical waves is

$$\varphi = \varphi_0^{(0)} - \alpha p(m)(1 - 1/s^{1/2}), \tag{4.64a}$$

$$M = \bar{M}_0 s^{-1/2}\left\{1 + \frac{\alpha}{2}q(m)(s^{1/2} - 1)s^{-1/2} - \right.$$

$$\left. - \frac{\alpha\bar{M}_0}{\lambda}(s^{1/2} - 1)\right\}, \tag{4.64b}$$

$$A = s - \alpha q(m)s^{1/2}(s^{1/2} - 1). \tag{4.64c}$$

Here $p = d\bar{M}_0/dm$, $q = d^2M_0/dm^2$ and the coordinate s increases from 1 to the infinity. Obviously, the small nonlinear distortions change the deviation angle and cause either partial focussing (Figure 4.1) or defocussing, dependent on the sign of the function q(m). If the amplitude at the front

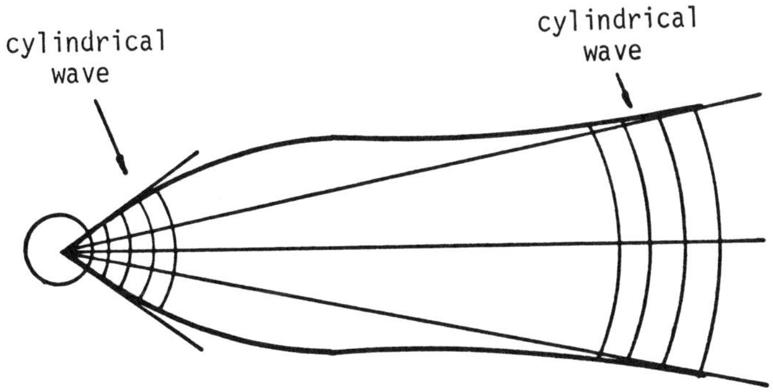

Figure 4.1. The nonlinear distortion of rays

undergoes a linear change ($p \neq 0$, $q = 0$), then in this approximation the focussing process does not occur. The exact solutions of nonlinear rays are given in [31, 85, 99], where the self-refraction of shock waves and solitons is investigated.

References

1. Ablowitz M.J., Benney D.J. The evolution of multiphase modes for nonlinear dispersive waves. - Stud. Appl. Math., 1970, v. 49, p. 225-238.
2. Asano N. Wave propagation in nonuniform media. - Suppl. Progr. Theor. Phys., 1974, No. 55, p. 52-79.
3. Asano N., Taniuti T. Reductive perturbation method for nonlinear wave propagation in inhomogeneous media. - J. Phys. Soc. Japan, 1969, v. 27, No. 6, p. 1059-1062.
4. Babich V.M., Buldyrev V.S. Asymptotic methods in the diffraction problems of short waves. - Nauka, Moscow, 1972 (in Russian).
5. Bagdoyev A.G. Wave propagation in continuous media. - Acad. Sci. Armenian SSR, Yerevan, 1981 (in Russian).
6. Bagdoyev A.G., Oganyan G.G. The propagation of nonlinear wavetrains in a relaxing gas-liquid medium. - Izv. AN SSR, Mekhanika zhidkosti i gaza, 1980, No. 1, p. 133-143 (in Russian).
7. Bakhvalov N.S., Zhileykin Ya.M., Zabolotskaya E.A. Nonlinear theory of wavebeams. Nauka, Moscow, 1982 (in Russian).
8. Blackstock D.T. Thermoviscous attenuation of plane, periodic, finite-amplitude sound waves. - J. Acoust. Soc. Amer., 1964, v. 36, p. 534-542.
9. Blackstock D.T. Connection between the Fay and Fubini solutions for plane sound waves of finite amplitude. - J. Acoust. Soc. Amer., 1966, v. 39, No. 6, p. 1019-1026.
10. Blatstein I.M. Calculation of underwater explosion pulses at caustics. - J. Acoust. Soc. Amer., 1971, v. 49, No. 5, p. 1568-1579.
11. Bogolyubov N.N., Mitropolsky Y.A. Asymptotic methods in the theory of nonlinear oscillations. - Gordon and Breach, New York, 1961.
12. Brekhovskikh L.M. Waves in layered media. - Nauka, Moscow, 1973 (in Russian).
13. Byatt-Smith J.B. An exact integral equation for steady surface waves. - Proc. Roy. Soc. London, 1970, A315, p. 405-418.
14. Byatt-Smith J.B. An integral equation for unsteady surface waves and a comment on the Boussinesq equation. - J. Fluid Mech., 1971, v. 49, p. 625-633.

15. Cole J.D. On a quasilinear parabolic equation occurring in aerodynamics. - Quart. Appl. Math., 1951, v. 9, No. 3, p. 225-236.
16. Cole J.D. Perturbation methods in applied mathematics. - Blaisdell, Waltham, MA, 1968.
17. Courant R. The partial differential equations. - Mir, Moscow, 1964 (in Russian).
18. Courant R., Friedrichs K. Supersonic flow and shock waves. - Inostrannaya Literatura, Moscow, 1950 (in Russian).
19. Crighton D.G. Model equations of nonlinear acoustics. - Ann. Rev. Fluid Mech., 1979, v. 11, p. 11-33.
20. Crighton D.G. Propagation of non-uniform shock waves over large distances. - In: Nonlinear Deformation Waves, IUTAM Symp., Tallinn, 1982; Springer, Berlin-Heidelberg, 1983, p. 115-130.
21. Crighton D.G., Scott J.F. Asymptotic solutions of model equations in nonlinear acoustics. - Phil. Trans. Roy. Soc. London, 1979, A292, p. 101-134.
22. Engelbrecht J. One-dimensional deformation waves in nonlinear visco-elastic media. - Wave Motion, 1979, v. 1, No. 1, p. 65-74.
23. Engelbrecht J. Two-dimensional nonlinear evolution equations: The derivation and the transient wave solutions. - Int. J. Nonlinear Mech., 1981, v. 16, No. 2, p. 199-212.
24. Engelbrecht J. Nonlinear wave processes of deformation in solids. Pitman, London, 1983.
25. Engelbrecht J.K., Nigul U.K. Nonlinear deformation waves. - Nauka, Moscow, 1981 (in Russian).
26. Erdelyi A., Bateman H. The integral transforms and the operation calculus. - Fizmatgiz, Moscow, 1961 (in Russian).
27. Fay R.D. Plane sound waves of finite amplitude. - J. Acoust. Soc. Amer., 1931, v. 3, p. 222.
28. Friedlander F.G. Sound pulses. - Cambridge University Press, 1958.
29. Fridman V.E. On propagation of intensive acoustic wave in the layered medium. - Akust. Zhurnal, 1976, v. 22, No. 4, p. 621-622 (in Russian).
30. Fridman V.E. Nonlinear acoustics of explosive waves. - In: Nonlinear Acoustics. Theoretical and Experimental Investigations. Gorki, Inst. Appl. Physics, Acad. Sci. USSR, 1980, p. 68-97 (in Russian).

31. Fridman V.E. Selfrefraction of small amplitude shock waves. - Wave Motion, 1982, v. 4, No. 4, p. 1-11.
32. Fridman V.E. Selfrefraction of weak shock waves. - Akust. Zhurnal, 1982, v. 28, No. 4, p. 551-559 (in Russian).
33. Fridman V.E., Petukhov Yu.V. Profile transformation of a propagating explosive wave. - J. Acoust. Soc. Amer., 1980, v. 67, No. 2, p. 702-704.
34. Fusco D., Engelbrecht J. The asymptotic analyses of nonlinear waves in rate-dependent media. - Il Nuovo Cimento, 1984, v. 80B, No. 1, p. 49-61.
35. Gaponov A.V., Ostrovski L.A. One-dimensional waves in nonlinear dispersive systems. - Izv. VUZ'ov. Radiofizika, 1970, v. 13, No. 2, p. 163-213 (in Russian).
36. Gasenko V.G. The map of the solutions to the Korteweg-de Vries - Burgers equation. - In: Investigations on hydrodynamics and heat transfer. Nauka, Novosibirsk, 1976, p. 81-87 (in Russian).
37. Germain P. Progressive waves. - In: Jahrb. DGLR (Köln), 1971, DGLR, Köln, 1972, p. 11-30.
38. Goldberg V.N., Zarnitsyna I.G., Fedoseyeva T.N., Fridman V.E. Relaxation effects in weak shock wave propagation in ocean. Akust. Zhurnal, 1981, v. 27, No. 1, p. 88-92 (in Russian).
39. Gorschkov K.A., Ostrovsky L.A., Pelinovsky E.N. Some problems of asymptotic theory of nonlinear waves. - Proc. IEEE, 1974, v. 62, No. 11, p. 1511-1517.
40. Grimshaw R. Slowly varying solitary waves. - Proc. Roy. Soc., London, 1979, A368, No. 1734, p. 359-388.
41. Grinfeld M.A. Ray method for calculating the wave-fronts in a nonlinear elastic medium. - Prikladnaya Matem. Mekh., 1978, v. 42, No. 5, p. 883-898 (in Russian).
42. Hopf E. The partial differential equation $u_t + uu_x = \mu u_{xx}$. - Communs Pure and Appl. Math., 1950, v. 3, No. 2, p. 201-230.
43. Ivanov I.D. Asymptotical description of a pulse reflecting from an interface. - Akust. Zhurnal, 1981, v. 27, No. 2, p. 234-242 (in Russian).
44. Jeffrey A., Engelbrecht J. Nonlinear dispersive waves in a relaxing medium. - Wave Motion, 1980, v. 2, No. 3, p. 255-266.

45. Jeffrey A., Kakutani T. Weak nonlinear dispersive waves: a discussion centered around the Korteweg-de Vries equation. - SIAM Rev., 1972, v. 14, p. 528-643.
46. Jeffrey A., Kawahara T. Asymptotic methods in nonlinear wave theory. - Pitman, London, 1982.
47. Joseph R.J. Solitary waves in a finite depth fluid. - J. Phys. A: 1977, v. 10, No. 12, p. 1225-1227.
48. Kadomtsev B.B., Petviashvili I.V. On the spectrum of the acoustical turbulence. - Doklady AN SSSR, 1970, 192, No. 4, 753-756 (in Russian).
49. Karpman V.I. Nonlinear waves in dispersive media. - Nauka, Moscow, 1973 (in Russian).
50. Kobelev Yu. A., Ostrovski L.A. Models of a gas-liquid mixture as a nonlinear dispersive medium. - In: Nonlinear Acoustics. Theoretical and Experimental Research. Acad. Sci. USSR, Gorki, 1980, p. 143-160 (in Russian).
51. Korn G., Korn T. Handbook of mathematics for scientists and engineers. - Nauka, Moscow, 1973 (in Russian).
52. Kravtsov Yu.A., Orlov Yu.I. Geometrical optics of inhomogeneous media. - Nauka, Moscow, 1980 (in Russian).
53. Kurin V.V., Nemtsov B.E., Eydman V.Ya. On the problem of sound beam reflection from the interface of two liquids. - Akust. Zhurnal, 1985, v. 31, No. 1, p. 62-68 (in Russian).
54. Landau L.D., Lifshits E.M. Mechanics of continuous media. - GITTL, Moscow, 1954 (in Russian).
55. Leibovich S., Seebass A.R. Nonlinear Waves. - Cornell University Press, Ithaca, N.Y., 1974.
56. Lerner A.M., Fridman V.E. Model equations of nonlinear acoustics in media with the high thermal conductivity. - Akust. Zhurnal, 1978, v. 24, No. 2, p. 228-237 (in Russian).
57. Levikov S.V. On transient weakly nonlinear internal waves in the deep ocean. - Okeanologiya, 1976, v. 16, No. 6, p. 968-974 (in Russian).
58. Maksimov B.N., Mikhailov G.D. The nonlinear equation of acoustics in the third approximation. - In: Nauchn. trudy. Moskovskiy univ. narodnogo khozyaistva, 1970, v. 96, p. 138-147 (in Russian).
59. Manukyan S.M. Numerical calculation of the nonlinear periodic shock wave in the vicinity of a caustic. - In: Uchenye zapiski Jerevanskogo universiteta, 1976, No. 2, p. 16-23 (in Russian).

60. Maslov V.P. The complex WKB-method for nonlinear equations. - Nauka, Moscow, 1977 (in Russian).
61. Miles I.W. Diffraction of solitary waves. - Z. angew. Math. and Phys., 1974, v. 25, No. 5, p. 889-902.
62. Miles, J.W. The Korteweg-de Vries equation: a historical essay. - J. Fluid Mech., 1981, v. 106, p. 131-147.
63. Mikhlin S.G. Variational methods in mathematical physics. - Nauka, Moscow, 1970 (in Russian).
64. Miropolski Yu.Z. Dynamics of internal gravitational waves in the ocean. - Gidrometeoizdat, Leningrad, 1981 (in Russian).
65. Mortell M.P., Seymour B.R. Pulse propagation in a nonlinear viscoelastic rod of finite length. - SIAM J. Appl. Math., 1972, v. 22, p. 209-224.
66. Mortell M.P., Varley E. Finite amplitude waves in bounded media: nonlinear free vibration of an elastic panel. - Proc. Roy. Soc. London, 1970, A318, No. 1533, p. 169-196.
67. Nariboli G.A., Sedov A. Burgers' - Korteweg-de Vries equation for viscoelastic rods and plates. - J. Math. Anal. Appl., 1970, v. 32, No. 3, 661-677.
68. Nakoryakov V.E., Shreiber I.R. A model of wave propagation in a gas-liquid mixture. - Teplofizika vysokikh temperatur, 1979, v. 17, No. 4, p. 798-803 (in Russian).
69. Naugolnykh K.A. The propagation of finite spherical sound waves in a dissipative medium. - Akust. Zhurnal, 1959, v. 5, No. 7, p. 80-84 (in Russian).
70. Naugolnykh K.A., Soluyan S.I., Khokhlov R.V. Cylindrical finite waves in a dissipative medium. - Vestnik MGU, Fizika, Astronomiya, 1962, No. 4, p. 65-71 (in Russian).
71. Naugolnykh K.A., Soluyan S.I., Khokhlov R.M. Spherical finite waves in the dissipative thermoconductive medium. - Akust. Zhurnal, 1963, v. 9, No. 1, p. 54-60 (in Russian).
72. Nayfeh A.H. Perturbation methods, John Wiley, New York, 1973.
73. Novikov A.A. On application of the method of coupled waves for non-resonant interactions. - Izv. VUZ'ov, Radiofizika, 1976, v. 19, No. 2, p. 321-328 (in Russian).

74. Ny A.L., Ryzhov O.S. Nonlinear wave propagation in media with the arbitrary number of chemical reactions. - Prikladnaya Matem. Mekh., 1976, v. 40, No. 4, p. 587-598 (in Russian).
75. Odulo A.B. On equations of long nonlinear waves in the ocean. - Okeanologiya, 1978, v. 18, No. 6, p. 965-971 (in Russian).
76. Oikawa M., Yajima N. Generalization of the reductive perturbation method to multiwave systems. - Suppl. Progr. Theor. Phys., 1974, No. 55, p. 36-51.
77. Ono H. Algebraic solitary waves in stratified fluids. - J. Phys. Soc. Japan, 1975, v. 39, No. 4, p. 1082-1091.
78. Ostrovski L.A. On wave theory in nonstationary compressible media. - Prikladnaya Matem. Mekh., 1963, v. 27, No. 5, p. 924-929 (in Russian).
79. Ostrovski L.A. Nonlinear internal waves in the rotating ocean. - Okeanologiya, 1978, v. 18, No. 2, p. 181-191 (in Russian).
80. Ostrovski L.A. Nonlinear internal waves in the ocean. - In: Nonlinear waves, Nauka, Moscow, 1979, p. 292-323 (in Russian).
81. Ostrovski L.A., Pelinovski E.N. Dynamics of nonlinear waves in a coastal zone. - In: Theory of waves, diffraction and propagation, VNIIRI, Jerevan, 1973, p. 342-349 (in Russian).
82. Ostrovski L.A., Pelinovski E.N. On the approximate equations for waves in weakly nonlinear dispersive media. - Prikladnaya Matem. Mekh., 1974, v. 38, No. 1, p. 121-124 (in Russian).
83. Ostrovski L.A., Pelinovski E.N., Fridman V.E. Propagation of finite acoustic waves in the inhomogeneous medium at the presence of caustics. - Akust. Zhurnal, 1976, v. 22, No. 6, p. 914-921 (in Russian).
84. Ostrovski L.A., Pelinovski E.N., Fridman V.E. Propagation of finite-amplitude acoustic waves in the stratified ocean. - In: Proc. VI Intern. Symp. Nonlinear Acoustics, Moscow University Press, 1976, v. 1, p. 342-353.
85. Ostrovski L.A., Shrira V.I. Instability and selfrefraction of solitons. - Zhurnal eksperimentalnoy i teoreticheskoy fiziki, 1976, v. 71, No. 4, p. 1412-1420 (in Russian).
86. Ostrovski L.A., Sutin A.M. Focusing of finite acoustic waves - Doklady AN SSSR, 1975, v. 221, No. 6, p. 1300-1303 (in Russian).
87. Ostrovski L.A., Sutin A.M. Diffraction and transfering acoustic N-waves. - Akust. Zhurnal, 1976, v. 22, No. 1, p. 93-100 (in Russian).

88. Pelinovski E.N. Nonlinear dynamics of tsunami waves. - Inst. Appl. Physics, Acad. Sci. USSR, Gorki, 1982 (in Russian).
89. Pelinovski E.N., Fridman V.E. Equations of nonlinear geometrical acoustics. - In: Nonlinear Deformation Waves, IUTAM Symp. Tallinn, 1982, Springer, Berlin-Heidelberg, 1983, p. 143-148.
90. Pelinovski E.N., Petukhov Yu.V., Fridman V.E. Approximate equations for the propagation of intensive acoustical signals in the ocean. - Izv. AN SSSR, Fizika atmosfery i okeana, 1979, v. 15, No. 4, p. 436-444 (in Russian).
91. Pelinovski E.N., Rabinovich M.I. On asymptotic method for weakly nonlinear distributed systems with variable parameters. - Izv. VUZ'ov, Radiofizika, 1971, v. 14, No. 9, p. 1374-1382 (in Russian).
92. Pelinovski E.N., Rayevski M., Schvaratski S.H. The Korteweg-de Vries equation for nonstationary internal waves in the inhomogeneous ocean. - Izv. AN SSSR, Fizika atmosfery i okeana, 1977, v. 12, No. 2, p. 325-328 (in Russian).
93. Pelinovski E.N., Soustova N.A. The structure of the nonlinear wavebeam in the inhomogeneous medium. - Akust. Zhurnal, 1979, v. 25, No. 4, p. 631-633 (in Russian).
94. Petviashvili V.I. Inhomogeneous solitons. - In: Nonlinear Waves, Nauka, Moscow, 1979, p. 5-19 (in Russian).
95. Romanov a R.R. On vertical propagation of short acoustic waves in the real atmosphere. - Izv. AN SSSR, Fizika atmosfery i okeana, 1979, v. 6, No. 2, p. 134-145 (in Russian).
96. Rozdestvenski B.L., Yanenko N.N. Systems of quasilinear equations and their applications in gas dynamics. - Nauka, Moscow, 1968 (in Russian).
97. Rudenko O.V., Soluyan S.I. Theoretical foundations of nonlinear acoustics. - Plenum Press, New York, 1977.
98. Sachs D.A., Silbiger A. Focusing and refraction of harmonic sound and transient pulses in stratified media. - J. Acoust. Soc. Amer., 1971, v. 49, No. 3, p. 824-840.
99. Schrira V.I. Nonlinear refraction of solitons. - Zhurnal eksperimentalnoy i teoreticheskoy fiziki, 1980, v. 79, No. 1, p. 87-98 (in Russian).
100. Scott A. Active and nonlinear wave propagation in electronics. - John Wiley, New York, e.a., 1970.

101. Sedov A., Nariboli G.A. Visco-elastic waves by the use of wavefront theory. - Int. J. Nonlinear. Mech., 1971, v. 6, p. 615-624.
102. Seymour B.R., Mortell M.P. Nonlinear geometrical acoustics. - Mechanics Today, 1975, v. 2, p. 251-312.
103. Shen M.C. Surface waves on viscous fluid of variable depth. - Phys. Fluids, 1976, v. 19, p. 1669-1675.
104. Shen M.C., Keller J.B. Ray method for nonlinear wave propagation in a rotating fluid of variable depth. - Phys. Fluids, 1973, v. 16, p. 1565-1572.
105. Sutin A.M. The influence of nonlinear effects on the properties of acoustical focusing systems. - Akust. Zhurnal, 1978, v. 24, No. 4, p. 593-601 (in Russian).
106. Taniuti T., Wei C.C. Reductive perturbation method in nonlinear wave propagation. I. - J. Phys. Soc. Japan, 1968, v. 24, p. 941-946.
107. Taniuti T. Reductive perturbation method on far fields of wave equations. - Suppl. Progr. Theor. Phys., 1974, No. 55, p. 1-35.
108. Taniuti T., Nishihara K. Nonlinear waves. - Pitman, London, 1983.
109. Tatsumi T., Tokunaga H. One-dimensional shock turbulence in a compressive fluid. - J. Fluid Mech., 1974, v. 65, pt 3, p. 581-601.
110. Theory of solitons. Method of the inverse problem, ed. Novikov S.P. Nauka, Moscow, 1980 (in Russian).
111. Towne D.N. Pulse shapes of totally reflected plane waves as a limiting case of the Cagniard solution for spherical waves. - J. Acoust. Soc. Amer., 1968, v. 44, No. 1, p. 77-83.
112. Van Dyke M. Perturbation methods in fluid dynamics. - Academic Press, New York, 1964.
113. Vasilyev A.B., Goldberg Z.A. Finite spherical waves in a viscous thermoconductive medium. - In: Proc. 6th All-union Acoustical Conf., 1968, Report B IV-I (in Russian).
114. Vinogradova M.B., Rudenko O.V., Sukhorukov A.P. Theory of waves. - Nauka, Moscow, 1979 (in Russian).
115. Whitham G.B. On the propagation of weak shock waves. - J. Fluid Mech., 1965, v. 1, No. 1, p. 290-318.
116. Whitham G.B. Linear and nonlinear waves. - John Wiley, New York, 1974.

117. Yeremenko V.A. On the velocity field structure in the vicinity of the caustic. - Prikladnaya Matem. Mekh., 1977, v. 41, No. 6, p. 1126-1130 (in Russian).
118. Yeremenko V.A., Ruzhov O.S. On flux structure at the neighborhood of shock wave intersection with the caustic. - Doklady AN SSSR, 1978, v. 238, No. 3, p. 541-544 (in Russian).
119. Yermakov S.A., Pelinovski E.N. On the role of nonlinear interactions in the forming of average fields. - Izv. AN SSSR, Fizika atmosfery i okeana, 1977, v. 13, No. 5, p. 537-542 (in Russian).
120. Zabolotskaya E.A., Khokhlov R.V. Quasisimple waves in nonlinear acoustics of bounded wavebeams. - Akust. Zhurnal, 1968, v. 15, No. 1, p. 40-47 (in Russian).
121. Zakharov V.E. Hamiltonian formalism for waves in nonlinear dispersive media. - Izv. VUZ'ov, Radiofizika, 1974, v. 17, No. 4, p. 431-453 (in Russian).
122. Zarembo L.K., Krasilnikov V.A. Introduction to nonlinear acoustics. - Nauka, Moscow, 1966 (in Russian).
123. Zeldovich Ya.B., Rayzer Yu.P. Physics of shock waves and high temperature hydrodynamic phenomena. - Nauka, Moscow, 1966 (in Russian).